Climate Policy and Development

Climate Policy and Development

Flexible Instruments and Developing Countries

Edited by

Axel Michaelowa

Head of Research Focus, 'International Climate Policy', Hamburg Institute for International Economics (HWWA), Germany

and

Michael Dutschke

Member of Research Focus, 'International Climate Policy', Hamburg Institute for International Economics (HWWA), Germany and Visiting Professor, University of São Paulo, Brazil

Edward Elgar

Cheltenham, UK • Northampton, MA, USA

Published by
Edward Elgar Publishing Limited
Glensanda House
Montpellier Parade
Cheltenham
Glos GL50 1UA
UK

Edward Elgar Publishing, Inc.
136 West Street
Suite 202
Northampton
Massachusetts 01060
USA

A catalogue record for this book
is available from the British Library

Library of Congress Cataloguing in Publication Data

Climate policy and development: flexible instruments and developing countries / edited by Axel Michaelowa and Michael Dutschke.
Includes bibliographical references.
1. Climatic changes—Government policy—Developing countries. 2. Greenhouse effect, Atmospheric—International cooperation. I. Michaelowa, Axel. II. Dutschke, Michael.

QC981.8.C5 C615 2000
363.738'747—dc21

99–087674

ISBN 1 84064 331 5

Printed and bound in Great Britain by MPG Books Ltd, Bodmin, Cornwall

Contents

Figures

Tables

Contributors

Regina Betz, Fraunhofer Institute for Systems Analysis and Innovation Research, Karlsruhe, Germany.

Preety Bhandari, Tata Energy Research Institute, New Delhi, India.

Michael Dutschke, Hamburg Institute for Economic Research, Hamburg, Germany and University of Sao Paolo, Brazil.

Wytze van der Gaast, Foundation Joint Implementation Network, Groningen, Netherlands.

Sandra Greiner, University of Hamburg, Hamburg, Germany.

Sujata Gupta, Tata Energy Research Institute, New Delhi, India.

Catrinus Jepma, University of Groningen, Groningen, Netherlands.

Karsten Krause, University for Economics and Policy, Hamburg, Germany.

Axel Michaelowa, Hamburg Institute for Economic Research, Hamburg, Germany.

Abbreviations and Acronyms

AE	Aeroenergia
AIJ	activities implemented jointly
BOT	build–operate–transfer
BPPT	Agency for the Assessment and Application of Technology
C	carbon
CACTU	Cantonal Agricultural Center of Turrialba
CAF	Certificados de Abonos Forestales
CATIE	Centro Agronómico Tropical de Investigacion y Enseñanza
CCB	Certificado de Conservación de Bosque
CDM	clean development mechanism
CER	Certified Emission Reduction
CNFL	Compania Naciónal de Fuerza y Luz
CO_2	carbon dioxide
COP	Conference of the Parties
CTO	certified tradable offset
DGEED	Directorate General of Electricity and Energy Development
ECOLAND	Esquinas Carbon Offset Land Conservation Initiative
EDF	Electricité de France
EI	Edison International
EPA	Environmental Protection Agency
ERU	Emission Reduction Unit
EU	European Union
FCCC	Framework Convention on Climate Change
FESP	Forestry Environmental Services Payment Programme
FOF	Fondo de Desarollo Forestal
FONAFIFO	Fondo Nacional de Financiamento Forestal
FSA	Financial Sustainability Analysis
GCA	Guanacaste Conservation Area
GEF	Global Environmental Facility
GHG	greenhouse gas
G77	Group of 77
HQ	Hydro Quebec
HS	hybrid system

IADB	Inter-American Development Bank
ICE	Instituto Costaricense de Electricidad
ICLEI	International Council on Local Environmental Initiatives
IEA	International Energy Agency
IET	International Emissions Trading
INC	Intergovernmental Negotiating Committee
IPCC	Intergovernmental Panel on Climate Change
IRR	Internal Route of Return
JI	Joint Implementation
KANSAI	Kansai Electric Power
KUD	Koperasi Unit Desa (village union)
LAPAN	National Institute of Aeronautics and Space
LOI	letter of intent
MHP	micro hydro power
MINAE	Ministerion de Ambiente y Energia
MME	Ministry of Mines and Energy
MOU	memorandum of understanding
MW	megawatt
NGO	non-governmental organization
NOx	nitrogen oxide
NTT	Nusa Tenggara Timur
O&M	Operation and Maintenance
OCIC	Oficina Costaricense de Implementacion Conjunta
ODA	Official Development Assistance
OECD	Organization of Economic Cooperation and Development
OH	Ontario Hydro
PAP	Protected Areas Project
PE	Plantas Eolicas
PFP	Private Forestry Project
PLD	Perusahaan Listrik Desa (village utility)
PLN	Perusahaan Listrik Negara (state electricity company), Indonesia
PPA	power purchase agreement
PV	photovoltaic
QELRC	Quantified Emission Limitation or Reduction Commitment
R&D	Research and Development
RESS	Renewable Energy Supply Systems
RET	Renewable Energy Technologies
SAR	Second Assessment Report
SBI	Subsidiary Body of Implementation

SBSTA	Subsidiary Body for Scientific and Technological Advice
SEI	socioeconomic integration
SELCO	Solar Electric Light Company
SGS	Société Générale de Surveillance
SHIS	Statistical Household Income Survey
SHS	solar home system
SINAC	Sistema Nacional de Aréas de Conservación
SO_2	sulphur dioxide
TEPCO	Tokyo's Electric Power Company
TM	Tierras Morenas
UNCTAD	United Nations Conference on Trade and Development
UNDP	United Nations Development Programme
UNFCCC	United Nations Framework Convention on Climate Change
UNIDO	United Nations Industrial Development Organization
USAID	United States Agency for International Development
USIJI	United States Initiative on Joint Implementation
V	volt
W	watt
WBCSD	World Business Council for Sustainable Development
WWF	Worldwide Fund for Nature

Preface

The role of developing countries in international climate policy is discussed intensely in the international climate negotiations. A focus of the international debate is whether or not industrialized countries should be allowed to fulfil at least a part of their domestic obligation to reduce greenhouse gas (GHG) emissions by taking corresponding measures in developing countries. At one extreme, this type of cooperation which has been labelled in many different guises is seen as a least-cost option to mitigate global climate change. At the same time it is expected to initiate important transfers of technology and capital to the developing world. Its opponents fear a new colonialist intrusion into the environmental space of the developing nations and a repetition of failures witnessed in development cooperation. Moreover, many decision makers and the public remain confused concerning the role of such cooperation in meeting the targets of the UN climate convention and the Kyoto Protocol.

This book features contributions from renowned specialists as well as from young German researchers that have not yet been available to an English-speaking audience. Chapter 1 outlines the evolution of the concept of flexible instruments emissions trading and joint implementation as well as the clean development mechanism (CDM) specially geared towards the developing countries since they were first proposed. It discusses their theoretical properties and the arguments of their proponents, as well as opponents. It summarizes the state of the activities implemented jointly (AIJ) pilot phase. Moreover; it outlines possible project categories and discusses who might be interested in using these mechanisms.

Chapter 2 analyses the decision on four different mechanisms laid down by the Kyoto Protocol as the result of diverging interests. The advantage of four over only one such mechanism lies in the fact that more interests could be considered. Where interests are strictly opposed and hard to reconcile, the creation of different systems seems the only way out. While the CDM as the relevant mechanism for the developing countries is very close to what politicians of host countries would consider optimal, placing emphasis on the developmental rather than on the environmental issue, the provisions about joint implementation between Annex I countries (industrialized countries with emission targets) seem to be more in line with investors' interests. Since the latter will favour the more liberal conditions of joint implementation with countries in transition, they might shy away from the CDM as long as marginal costs of abatement measures are not considerably lower in the developing countries and as long as there is no shortage of project opportunities in the former countries. As some developing countries might fear being cut off from the benefits of additional transfers, they might change sides and adopt national targets.

Chapter 3 addresses questions related to the CDM:

1. How to determine the GHG emission reductions achieved through a CDM project? Not only should the emissions of the new plant under the project be measured, but also a reference scenario ('baseline') needs to be determined showing what the emissions would have amounted to in the absence of the project.
2. Are forestry projects allowed under the CDM? In its current form, Article12 of the Kyoto Protocol only refers to projects that result in emission reductions, but some observers have argued that this does not necessarily exclude 'sinks' projects.
3. What will the governance structure of the CDM look like? The Executive Board for the CDM could be assisted by so-called 'operational entities', but it is not yet clear which institute will do what and who will be responsible for certification procedures.
4. What exactly will be the amount that parties involved in CDM projects have to pay to cover administrative expenses and adaptation investments in the developing countries most vulnerable to climate change?

Chapter 4 discusses ways and means for a country to elicit CDM funding. Transparency in the project selection and approval process is of utmost importance. The foremost criterion for the host country is to undertake a national exercise involving the private sector in prioritizing the sectors where such funding and projects should be undertaken, which should obviously be in line with the development objectives of the country. To alleviate distrust of developing countries an experimental phase should be set up.

Chapter 5 is the first of two case studies, which evaluates a German/Japanese AIJ project to disseminate solar home systems and small-scale renewable energy supply systems in Indonesia. It discusses the large number of objectives that should be achieved within a CDM project to ensure that it fulfils both the climate change (CDM, primary objectives) and development-related criteria. The analysis reveals a number of problems such as life-cycle emissions, leakage effects, transaction costs, delays in the time schedule, double counting and the importance of proper reporting.

The second case study (Chapter 6) discusses Costa Rican AIJ policy, one of the rare instances where the process is host country-driven. From the very beginning, Costa Rica has understood climate cooperation as one way of financing its development goals. It has been active in the AIJ process and in the UN Framework Convention on Climate Change (UNFCCC) climate negotiations. The chapter undertakes an evaluation of the way project cooperation developed within the period 1994/1997 and gives a complete overview of all projects actually approved in the small central American country, including the calculation of climate benefits and externalities, and how the programme was developed by the hosts later on.

Chapter 7 stresses the role of renewable energy technologies in the CDM. It becomes clear that two factors are crucial for the implementation of a sustainable energy system. First is the balancing of the commercial incentives of private sector actors with the CDM requirement of additional projects that would have happened anyway; second is the political recognition of energy as a basic need, including the individual right to energy access that is required for clean water and food. This right includes the physical and economic access to energy.

Chapter 8 sums up the problems the forthcoming negotiation rounds will have to solve and proposes practical solutions for the design of the CDM, taking into account the results of the 5th conference of the parties at Buenos Aires in 1998. It discusses the process of setting baselines, the institutional structure and credit sharing. Moreover, a more long-term view of the interaction of climate and development policy is taken, aiming at an equitable allocation of emission rights.

AXEL MICHAELOWA
MICHAEL DUTSCHKE

1. Flexible Instruments of Climate Policy

Axel Michaelowa[1]

FUNDAMENTAL ASPECTS OF INTERNATIONAL CLIMATE POLICY

Anthropogenic Climate Change

Through industrialization, increasing consumption of fossil fuels and changes in land use, man has a considerable influence on the composition of the earth's atmosphere. Increases in the concentration of source gases such as carbon dioxide (CO_2), methane (CH_4), nitrogen oxide (NO_x) and chlorofluorocarbons (CFCs) are in some cases dramatic. Many of these gases allow the short wave rays of the sun to reach the earth's surface unhindered while warming the atmosphere.[2] Computerized climate models have projected that a doubling of equivalent CO_2 concentrations, which will be reached by 2030 if emission trends are extrapolated, will lead to a global temperature increase of 1°C ± 0.5°C by 2030 and of 3°C by 2100 (IPCC, 1996). This temperature increase corresponds to the warming which took place between the last ice age and the current warm epoch. The faster it occurs, the greater will be its effect on the environment. Ecosystems can only adapt to the change if the rate of temperature increase is reduced. Although there is still considerable uncertainty concerning the regional consequences of climate change, it must be assumed that they will tend to be negative and, in some cases, of disastrous dimensions. Moreover, it is likely that damage will not be equally distributed across the globe. A major assessment of regional impacts

1. I thank Kai-Uwe Schmidt of the UNFCCC secretariat for providing me with latest data about the number and type of approved AIJ projects.
2. Even if there are still a few scientists who dispute this, the following statement from Professor Graßl, director of the Max-Planck-Institute for Meteorology in Hamburg, illustrates the strong scientific consensus regarding the emergence of climatic change: 'As far as the scientific side is concerned, the circumstantial evidence in the case of the ozone layer and the greenhouse effect has been examined, the verdict of guilty has been passed and documented in the files of the United Nations. It is no longer a question of collecting more proof, but solely of reducing the "sentence" at least as far as the climate is concerned' (quoted in Frese, 1994, pp.5f).

of climate change (IPCC, 1997) concluded that developing countries will be more vulnerable than industrial ones. While agricultural yields will probably increase in middle and high latitudes, yields in the tropics are likely to fall. A rise in the sea level will have its major impact on the small, low-lying island states of the tropical oceans. Rainfall variability is likely to increase, affecting marginal communities. The range of tropical diseases will expand. Inter alia, the IPCC concludes that 'the African continent is particularly vulnerable. ... The human, infrastructural and economic response capacity to effect timely response action may well be beyond the economic means of some countries' (ibid., p.9).

The Need for a Global Approach

To stabilize greenhouse gas concentrations at the current level, an immediate emission reduction of at least 60 per cent would be necessary (Grubb and Rose, 1992, p.3). However, since many greenhouse gases cause no direct environmental damage on a local or regional level, and since the earth's atmosphere is a public good which can be used at no cost by all nations for the absorption of greenhouse gases, there has so far been little incentive to reduce emissions.[3] The effectiveness of isolated national policies is also very limited. Owing to the fact that possible damage appears on a global level and only after a certain time lag, it is impossible to apply the 'polluter pays' principle. Moreover, it is likely that damage will fall mainly on those countries that have not contributed to greenhouse gas emissions themselves. Free-riding by individual nations which fail to implement their own policies to reduce climate change yet benefit from measures taken by other states cannot be ruled out. Climate policy therefore requires international coordination.

Equity

Equity considerations are generally held to be an important aspect of climate policy and of achieving sustainable development.[4] So far, they have been invoked very often, but not really cared about. Equity involves procedural as well as consequential issues. Procedural issues relate to the way decisions are made, while consequential issues relate to outcomes. To be effective and to

3. The price of fossil fuels provides incentives to lower consumption and so reduce emissions. These incentives, however, do not correlate with the emission intensity of the energy carriers and are, therefore, unsystematic. In addition to that, prices of fossil fuels have been falling in the past 30 years, owing to the worsening of the terms of trade for the producing countries.
4. In common language, equity means 'the quality of being impartial' or 'something that is fair and just'.

promote cooperation, agreements must be regarded as legitimate, and equity is an important element in gaining legitimacy.

Procedural equity encompasses process and participation issues. It requires that all countries be able to participate effectively in international negotiations related to climate change. Appropriate measures to enable developing countries to participate effectively in negotiations increase the prospects for achieving effective, lasting and equitable agreements on how best to address the threat of climate change. Concern about equity and social impacts points to the need to build endogenous capabilities and strengthen institutional capacities, particularly in developing countries, to make and implement collective decisions in a legitimate and equitable manner. This also applies to participation of non-governmental organizations (NGOs).

Consequential equity has two components: the distribution of the costs of damages or adaptation and of measures to mitigate climate change. Climate change impacts will be distributed unevenly. Because countries differ substantially in vulnerability, wealth, capacity, resource endowments and other factors listed below, unless addressed explicitly, the costs of the damages, adaptation and mitigation will be borne inequitably. As mentioned above, it is likely that industrial countries will be able to adapt at comparatively low cost, whereas developing countries struggle with adaptation. The latter often have different urgent priorities and weaker institutions, and are generally more vulnerable to climate change.

There are substantial variations both among developed and developing countries that are relevant to the application of equity principles to emission reduction. These include variations in historical and cumulative emissions, current total and per capita emissions, emission intensities and economic output, projections of future emissions and factors such as wealth, energy structures and resource endowments.

A variety of ethical principles, including the importance of meeting people's basic needs, may be relevant to addressing climate change, but the application of principles developed to guide individual behaviour to relations among states is complex and not straightforward. Climate change policies should not aggravate existing disparities between one region and another, nor attempt to redress all equity issues.

It is necessary that both efficiency and equity concerns be considered during the analysis of emission reduction and adaptation measures and instruments. It should be possible to combine equity and efficiency.

The International Framework of Climate Policy

International coordination of climate policy has happened increasingly over the past few years, even if we are still far from even bending the growth curve of global greenhouse gas emissions. The Framework Convention on Climate Change, which was passed during the 1992 UN Conference on Environment and Development at Rio de Janeiro, was very vague as far as reduction targets and financial commitments are concerned. In the Convention, the industrialized nations made a non-binding commitment to reduce their greenhouse gas emissions to 1990 levels by 2000.

There were no such commitments for the developing countries owing to the principle of 'common, but differentiated, responsibilities' (Art. 3.1). The Convention's preamble explicitly states the need for economic development, that is economic growth and elimination of poverty. If these aims were achieved, the share of emissions of developing countries would grow. Developing countries agreed to set up emissions inventories financed by the industrialized countries (Art. 12). They expected the latter to pay the 'agreed full incremental costs' of emission reduction measures (Art. 4.3) and fund adaptation measures (Art. 4.4). Moreover, technology transfer was to be promoted (Art. 4.5). The hope of developing countries that they would get substantial amounts of funding through these provisions quickly proved too optimistic. Instead of the hoped-for $140 billion for the period until 2000 that had been mentioned in Agenda 21, only a trickle of money flowed. The Global Environment Facility (GEF) that finances projects under the climate and biodiversity convention as well as under the Montreal Protocol disbursed just $733 million in the period 1991–1994. Up to mid-1997, a further $883 million was allocated (Porter *et al.*, 1997), of which 45 per cent went into climate change projects, many of which were linked to the setting-up of national inventories. Overall funds for GEF stand at $2.75 billion for 1998–2002 (Watson, 1998). The meagre funding led developing countries to vigorously oppose pressures from some industrial countries, particularly the USA, to commit themselves to emission targets.

The Convention entered into force in 1994. At the first conference of the parties in Berlin in 1995, negotiators agreed to negotiate a protocol with legally binding emission targets until 1997. The Kyoto Conference in 1997 achieved this aim: the Kyoto Protocol set binding targets for the industrial countries while developing countries once more failed to subscribe to targets. They even deleted a draft article which would have allowed voluntary take-up of targets. Many details, such as the inclusion of sinks were left open, though, owing to the time constraints in the final days of negotiations. The Buenos Aires Conference of the Parties in 1998 decided on a work programme to resolve issues of the Kyoto Protocol by 2000.

From the beginnings of climate policy, a major issue has been how it can be implemented cost-effectively in a world where only some countries have an emission target. The Convention explicitly states this goal in Art. 3.3, giving on to say that climate policy should 'ensure global benefits at the lowest possible cost'. This in practice means that, even if emission targets apply only to a small subset of countries, all possibilities worldwide have to be used to reach these targets. How to tap this potential has been the subject of a long and often bitter discussion which will be surveyed in this chapter. The instruments finally agreed in Kyoto have been called 'flexibility instruments'.

FLEXIBILITY INSTRUMENTS: ECONOMIC THEORY

A climate policy instrument makes sense economically when it leads to a reduction in emissions such that the benefit and cost of an additional unit of emission reduction (marginal benefit and marginal reduction cost) are equal. Since it is impossible to determine the marginal benefit in reality because of long lags and uncertain impacts, an approach must be pursued whereby targets are the result of a political decision. Whether or not these targets are acceptable from an environmental point of view depends upon the political assertiveness of environmental interests. The task of achieving optimal results then lies in reaching the set targets at minimum cost, as has been enshrined in the Convention. This approach consequently requires the development of instruments which lead people and companies to continue reducing their emissions until their marginal costs are equal. The evaluation of climate policy instruments should not, however, be limited to economical efficiency but should also take into consideration the extent to which the environmental goal has been reached, as well as the effects on innovation, and the administrative and legal aspects of its implementation.

The global applicability of the instruments considered below is examined under the assumption that conditions remain unchanged and that the instruments are implemented in an ideal form. The evaluation of taxation and emission rights entails a high degree of abstraction. The analysis is valid only for the effects of each individual instrument. It is to be expected that, in reality, none of these instruments will be implemented in isolation, but rather as component parts of a bundle of interacting instruments, since this is easier to carry through politically and because in this way other targets, such as fair allocation, can also be achieved. An analysis of specific possible bundles of instruments is not carried out here.

Global Greenhouse Gas Tax

A global greenhouse gas tax would provide incentives to lower emissions because reduction measures are worthwhile when they cost less than the tax load on the reduced emissions. However, besides the issue of sovereignty, such a tax would present considerable conceptual and allocational problems. Since products and services whose production leads to greenhouse gas emissions would become more expensive, demand would fall as long as demand price elasticities were within normal limits. The competitive disadvantages and displacement effects (Hoel, 1993) resulting from the price rises feared in the case of unilateral action would not materialize in the case of a global solution. Owing to the orientation of international climate policy on absolute emission targets, the tax is not attractive as a target cannot be achieved *exactly* by means of the tax. Tax-induced emission reductions depend upon a number of economic factors – such as the price elasticities mentioned above – which are unstable and subject to unpredictable change. Given an overachievement of the reduction target in the case of an excessive rate of taxation, the cost to the economy of levying the tax would grow, since the marginal reduction costs rise with increasing reduction. If the rate of taxation were too low, the reduction target would not be reached and the tax rate would have to be adjusted upwards. Fine-tuning the rate of taxation would thus be a necessary and highly complex task. Even if it were possible to come to a consensus on a particular rate of taxation, further problems would emerge. It is debatable whether exchange rates or purchasing power parities should be used in order to convert the taxation rate into national currencies. Existing taxes, such as petrol tax, and subsidies would need to be taken into account when determining the national rate of taxation and would lead to various national rates of emission tax if a standard global rate of taxation were to be established.[5]

The tax revenue could be used either to lower budget deficits, to reduce distortions in the existing taxation system, or to finance emission reduction projects. In order to reach a specific reduction target, the first two possibilities require a higher rate of taxation than the third. Dismantling tax system distortions of a primarily political nature would reduce the economy-wide cost of the tax. The emission tax could thus be an instrument for reducing political distortions of the system. When it comes to distributing tax revenues, however, there is a danger that the interests of well-organized lobby groups representing emission-intensive branches of industry could influence the behaviour of state bureaucracies. Economically undesirable competitive imbalances in favour of powerful interest groups could result. The term 'revenue neutrality' is often mentioned in this context. Revenue neutrality really means that all state revenue generated by the emission tax should be reimbursed to the taxpayers as a whole. Interest groups have a different

5. Let Country A levy an oil tax of US$2/t C (ton of Carbon) while Country B levies no oil tax. In order to arrive at a global tax rate of US$10/t C, an emission tax of US$8/t C is required in Country A and US$10/t C in Country B.

interpretation. Each interest group demands revenue neutrality for itself.[6] Granting this would remove the incentive to reduce emissions and dilute the efficiency of the tax.

Developing countries would be afraid that their few, mostly emission-intensive, industries would lose competitiveness and so demand compensation payments. Furthermore, exporters of fossil fuels would come out in favour of instruments which maintain their incomes. This would be the case with export taxes. International agreement on the distribution of tax revenues seems hardly possible.

In contrast to other instruments, the emission tax honours previous efforts made towards emission reductions. Manufacturers who reduced their emissions before introduction of the tax benefit from low tax payments, while manufacturers with high emission levels have a heavy burden to bear.

The debate surrounding the introduction of a CO_2/energy tax in the EU clearly demonstrates the problems of the tax solution. The current proposal, which has so far failed to find a consensus, would lead to enormous distortions owing to numerous special arrangements for energy-intensive industries and entire countries. From the pure point of view of climate policy it is, for example, completely incomprehensible why electricity from nuclear power stations should be subject to taxation since its generation causes virtually no greenhouse gas emissions. Taxing nuclear energy can therefore only be justified by other externalities such as radiation risks. However, it would require a highly complex arrangement of environmental taxes if all the externalities involved in energy production were to be taken into consideration. A general energy tax does not solve this problem since it cannot take sufficient account of the various externalities.

Given the numerous conceptional problems mentioned above and the course of international climate negotiations which early on took the decision to concentrate on emission quantities, not instruments, it is extremely unlikely that a global greenhouse gas tax will materialize in the near future. It may well be however, that the idea will resurface in the long term, possibly as a consequence of disappointments with emissions trading and a resurgence of a more interventionist approach that needs fiscal revenues. The current insistence of the EU on harmonized policies and measures could be a first step in this direction.

---

6. For example, an energy-intensive branch of industry demands a complete tax reimbursement for reasons of 'competitiveness'. In the case of revenue neutrality for the economy as a whole, this industry's tax burden would be higher than the reimbursement, if the reimbursement was to take place without distortion.

Emission Trading

Given an emissions budget, perfect markets and harmonized standards of emissions monitoring, global emission trading represents the most efficient way of achieving a stipulated level of emissions. In this quantitative solution, the price per ton of emissions is not fixed by the state as in the case of a tax but, with a fixed limit on total emissions, is set by the market. Governments can trade with their emissions allocation. States with reduced emissions can sell emission rights they do not need, thus benefiting from the reduction achieved. Countries that have to purchase additional emission rights in order to maintain their production or consumption also have an incentive to reduce emissions. Emission rights can also be banked for use at a later date. Thus states adjust their emissions in such a way that the marginal reduction costs correspond to the price of the emission rights. In this way, emission reductions are achieved in a cost-efficient manner, especially if private actors can participate in trading. The extent of a possible increase in efficiency becomes clear from studies which calculate cost savings by making the transition from national to international emission targets. The 1997 comparison of seven economic models by the Stanford Energy Modelling Forum calculates cost savings of 20–90 per cent through global emission trade. The initial allocation of emission permits at the international and national level poses many problems. Related problems at an international level are discussed extensively by Grubb *et al.* (1992; see also Rose *et al.*, 1998).

Developing countries demand the proportional allocation of emission rights to each country according to the size of its population. However, some authors (Grubb *et al.*, 1992) have suggested that this form of allocation should not provide an incentive for increasing the population growth rate and should, therefore, be based on a particular year in the past. However, this proposal would throw up massive allocational problems, as would the proposal to allocate emission rights in inverse proportion to per capita GNP (Rose, 1992, p.75). Countries with high per capita emission levels would have to purchase large quantities of emission rights while countries with low emissions could siphon off high profits. The result would be a global redistribution process comparable to that witnessed during the oil crises from which the OPEC states benefited. One possible compromise solution would be to issue initial emission rights according to previous emissions, with new issues, after a transition period, being dependent upon the population size in a base year (Sanhueza *et al.*, 1994, p.14).

On a national level, established producers and consumers benefit from a cost-free initial issue (grandfathering) which corresponds to their previous emissions, while newcomers must always purchase emission permits.

Moreover, producers who already work efficiently are disadvantaged because their previous efforts to reduce emissions are not taken into account. Some of these points of criticism can be taken into consideration by auctioning the emission permits.

Until now, only the USA has had experience of tradable, limited-period emission permits (for example, for SO_2) over a relatively long period of time which has confirmed in principle the feasibility of this instrument and demonstrates a great potential for increasing efficiency.[7] Transferring this instrument to greenhouse gases presents no problem. In the USA, most transactions took place within companies with several production plants. This implies relatively high transaction costs, which are estimated at 10–30 per cent of total costs. Barrett, on the other hand, mentions transaction costs of less than 1 per cent (Barrett, 1994, p.12). The discrepancy between these results demonstrates that there is great uncertainty regarding the level of transaction costs, and that further research is necessary.

The development of efficient markets for emission rights is hampered if only governments can trade or if, at a national level, the number of potential dealers is small. Under the former limitation no liquid market will come into being: there will be a series of bilateral transactions without transparent prices. Furthermore, excessive regulation of individual transactions and the fear of future changes to the system or to the underlying environmental standards can restrict trade (ibid., p.13).

The participation of the private sector complicates the system as the national instruments of climate policy have to be taken into account. Efficiency of international emission trade can be further enhanced if these instruments are also efficient. Emission targets have to be set in such a way that they are reachable. Their updating and strengthening in future commitment periods has to be foreseeable and should then actually take place. Sanctions that are applied in the case of violation of rules have to be reliable.

---

7. In 1976, the US Environmental Protection Agency (EPA) introduced the possiblity of compensating for exceeding emission standards at some plants with emission reductions achieved at other plants. By 1992, there had been 2500 compensations. In 1979, the bubble principle was introduced which fixes an aggregate emissions target for all the emitters in a region. Emission rights can be traded within that region. It is also possible to 'save' emission rights. The emission rights approach was particularly successful during the transition to unleaded petrol and in the case of SO_2 emissions from power stations (Harper, 1994, pp.133ff). The latest programme, which was anchored in the 1990 Clean Air Act, began on 1 January 1995 and allows for emission rights which was traded between about 100 power stations, to be extended to 900 power stations as from the year 2000. Emission rights were traded as early as 1992 on the Chicago futures exchange. Any emissions not covered by emission rights are fined at a rate of US$2000 per ton. The success depends decisively on rigorous verification (ibid., pp.139f).

Given the extreme differences between the political interests of the nation states, it was long seen as unlikely that a global system of tradable emission rights for greenhouse gases would emerge in the near future. This changed with the Kyoto Protocol, which allows emission trading between countries with targets in principle. Negotiations in Buenos Aires have shown that the question of initial allocation is still so potentially explosive that a consensus is difficult to achieve. Agreement on the political and legal framework (for example, stock exchange supervision) and on the necessary international sanctions and verification mechanisms also presents difficulties. As a transitional solution, a medium-term model is conceivable which, in line with the Montreal Protocol, allows for a system of emission trading in the industrialized countries.

Reaching Targets through Projects Abroad: Joint Implementation and the Clean Development Mechanism

The meaning of the term 'Joint Implementation' (JI) has changed several times during climate negotiations. In the beginning it was used for any type of instrument that can be applied globally, and thus had the same meaning as 'flexibile instruments' today. After Rio, it took on the meaning that countries may invest in projects abroad and get the emission reduction credited to their domestic target. At Kyoto, its meaning was further narrowed by adding the restriction that all participating countries have to have targets. The Kyoto Protocol allows projects in countries without targets under the so-called 'Clean Development Mechanism' (CDM). Currently, this is the only flexibility instrument involving developing countries.

JI and CDM do not lead to the allocation problems associated with both the greenhouse gas tax and emission rights, yet, compared to a global greenhouse gas tax or a global system of emission trading, they are only a 'second best' solution. A necessary condition for JI and CDM is the existence of national emission reduction targets in the investor country which are to be achieved by means of taxation or regulation instruments.[8] This is precisely the situation in which many industrialized countries find themselves today. JI and CDM take place on two levels. The principle is illustrated in Figure 1.1.

---

8. Voluntary commitments by companies are also conceivable. However, these can only work if the state threatens convincingly to introduce taxation or regulations in case the voluntary commitments are not met. Otherwise companies would have no incentive to actually comply with their voluntary commitments. In the following, voluntary commitments are therefore subsumed under taxation and regulation.

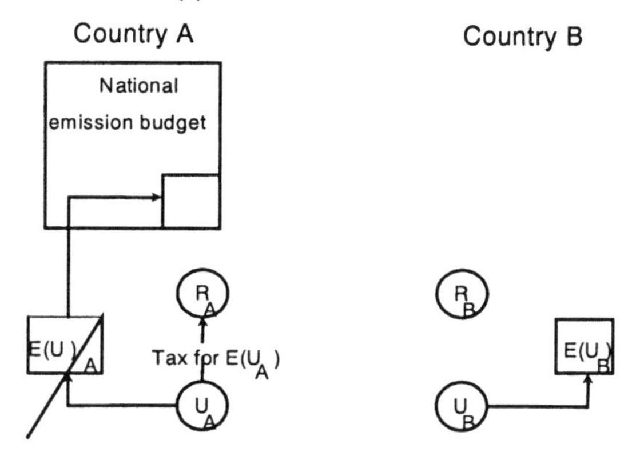

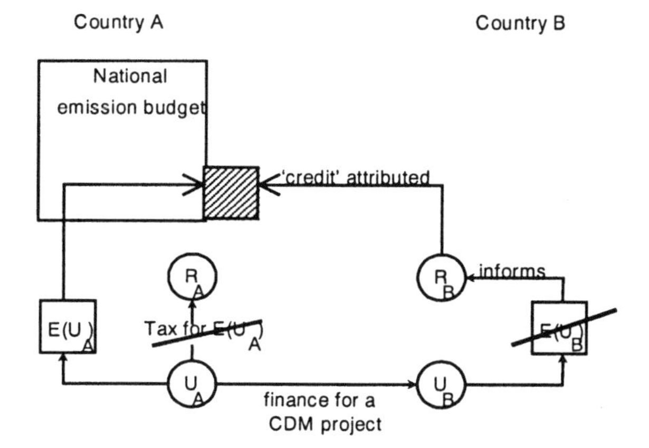

Figure 1.1 *How the CDM works*

Company U_A pays the government R_A which is committed to a national reduction target, an emission tax on its emissions $E(U_A)$. Company U_B emits a volume of greenhouse gases $E(U_B)$ without any restrictions from its government R_B. Let $E(U_A) = E(U_B)$ in this example. The reduction project financed by U_A completely eliminates U_B's emissions $E(U_B)$. The government R_B is informed of the reduction which is then attributed to Country A's emission budget. The government R_A grants U_A tax concessions which, in the example, correspond exactly to the tax burden created by $E(U_A)$, since $E(U_A) = E(U_B)$. Total emissions are thus reduced by 50 per cent.[9]

9. U_A could change its behaviour, leading to an increase in $E(U_A)$. It must, however, pay tax on the additional emissions as before. In this way, there is no incentive to increase emissions in the home country.

Under JI the host country would also have a budget and substract the credited amount from its budget. First, reductions achieved abroad are credited to the investor country's reduction target. Second, domestic emitters who can prove reductions abroad are granted tax relief or are subject to fewer regulations. Such concessions are proportional to the reduction achieved. If an emission tax is in force, the concession will be equal to the tax payments that will have to be made for corresponding emissions at home. Domestic emitters will continue to finance projects abroad until marginal reduction costs are equal at home and abroad.

Given sufficiently large marginal reduction cost differences between the investor country and abroad, there exists an incentive to *fully* finance projects in other countries whose sole aim is the reduction of greenhouse gas emissions. If the differences in marginal reduction costs are not large enough for this purpose, there is an incentive to carry out projects where the reduction of greenhouse gas emissions is just one among many aims or is even a not explicitly intended side-effect. Even with identical marginal reduction cost curves there are incentives for JI and CDM.

The advantage of JI and CDM lies in their flexibility. Not only the state, but also companies, NGOs or households can participate. Any project should be taken into consideration which is intended, either entirely or partially, to reduce greenhouse gas emissions. This contributes to a voluntary transfer of resources to the developing countries and the countries in transition. These solutions also do without a global reduction target; a national target in one of the participating countries suffices. JI and CDM can alleviate the competitive imbalances feared by industry in the case of unilateral national taxation or regulatory measures as part of that country's climate policy. Emission-intensive industries can reduce their taxation or regulatory burden by means of investing in projects abroad. The shock of a structural change required by environmental policy is thus cushioned. Short-term frictional costs, which could otherwise lead to social tensions, are reduced. Fewer job losses in the emission-intensive industries can be assumed, while job gains in industries which offer energy-efficient technologies can be expected.

Designing the framework for JI and CDM projects is very important because it has to provide the right incentives and keep transaction costs[10] to a minimum. Complete realization of potential efficiency gains is only possible if as many as possible of the economic agents who can influence greenhouse gas emissions have the opportunity to participate in projects. Administrative flexibility is a necessary condition for successful JI and CDM. Comparative studies in fields where JI is already possible at a national level (for example, SO_2 in the USA) have shown that efficiency gains are very modest when the

10. Transaction costs include the cost of information procurement, contract negotiations and verification.

concept is handled restrictively, whereas a flexible approach results in considerable improvements. The level of transaction costs is particularly dependent upon the projects' institutional framework, whereby it is necessary to examine the advantages and disadvantages of centralized and decentralized organizational forms. Centralization probably leads to higher costs but can also assist efficiency improvements and learning effects. For this reason there ought to be competition between various institutions for JI and CDM.

Furthermore, the degree to which this instrument can meet set targets is particularly dependent upon the choice of project-related criteria and the choice of project partners. An important role is also played by the existing reduction cost potentials and political circumstances in the relevant countries. A stable political environment is important as projects will often run for long periods of time. Investors can insure themselves to a certain extent by providing initial capital and making further payments dependent upon progress being made.

The institutional organization of JI and CDM and the quality of project monitoring, evaluation, verification and certification (see Table 1.1) have a crucial influence on the ability to attribute emission reductions as well as on a project's chances of success. Problems regarding the achievement of emission targets can result from possible collusion, since all the participants have an interest in declaring excessive emission reductions (Germanwatch, 1994, p.6). Thus the definition of the project baseline, that is the situation if the project had not taken place, is crucial to avoid the crediting of fictitious reductions. The baseline issue will be discussed in more detail in the case studies and in the final chapter. Equally, double counting of emission reductions must be avoided.

It is evident that, owing to information procurement costs, control costs and the costs involved in the complex negotiation of contract terms, potential projects with a modest absolute reduction cost advantage will not take place. Certain savings potentials must therefore be consciously sacrificed in order to prevent misuse of the JI and CDM and so ensure their long-term operability.

Table 1.1 Institutional options for JI and CDM projects

Institution	Function	Activity
Project exchange	To identify potential project participants	Makes it easier for investors and project instigators to contact each other
Broker	To help effectuate a project	Brings investors and project proponents together; tries to effectuate efficient projects
Consultant	To support project development	Offers technical, legal, engineering, financial and management advice
Approval authority	To accept the project	Determines whether the project conforms to international criteria
Financial intermediary	To ease finance	Offers projects as investment opportunities; bundles projects and sells shares in project bundles
Insurance	To safeguard the reduction	Bears the losses sustained by the government and the investor in the case of project failure
Certifier	To verify and certify the reduction	Verifies the reduction and issues confirmation (certificate)
Dispute settlement authority	To mediate in disputes and impose sanctions	Conducts conciliation procedures; decides who is responsible for the failure of a project and imposes sanctions

Source: Wexler *et al.* (1995, p.116); own additions.

Reaching Targets Together: Bubbles

Bubbles are the most far-reaching flexibility instrument. They come into being when countries adopt joint targets. Bubbles minimize transactions cost. Policies and measures in the bubble nevertheless have to be designed in such a way that they achieve emission reduction at minimum cost. This presumes a minimum of harmonization of policies or the possibility of internal emission trade. Bubbles may set differentiated targets for participating states as long as the overall target will not be surpassed.

NEGOTIATING HISTORY OF FLEXIBLE INSTRUMENTS

Flexible instruments were first discussed within the framework of international climate policy negotiations in 1989. At the Noordwijk (Netherlands) conference, the management consultants McKinsey presented a concept which envisaged the implementation of emission reduction projects via an international 'clearing house' (McKinsey and Company, 1989). Also in 1989, Michael Grubb from the Royal Institute for International Affairs, London, discussed international emission trading (Grubb, 1989).

However, these ideas took a back seat in the following period, although the McKinsey concept was mentioned by the USA, in spring 1991, in the Intergovernmental Negotiating Committee (INC), as part of a draft Convention text for the UN Conference on Environment and Development. It was not until the end of 1991 that Norway actively introduced it into the negotiations under the title 'Joint Implementation'. Norway was supported by Germany, which simultaneously made an issue of JI in the Federal Government's Third Resolution on CO_2 reduction of 12 December 1991, thus taking up, among others, ideas from the Netherlands and a proposal from the Federation of Germany Industry (BDI, 1991).

At the conference itself, however, the concept of JI stayed in the background, while emission trading was regarded as utopian and was not even discussed. JI was included in the Framework Convention on Climate Change without debate, and roughly outlined in Article 4.2. This was owing to another strand of the negotiation which later became known as 'the EU bubble'. The EU wanted to be able to reach a joint target. Article 4.2(a) states that parties to the Convention which are industrialized countries (OECD countries) and the other parties listed in Appendix I (countries in transition in eastern Europe and Russia) may, together with other parties, carry out measures for limiting greenhouse gas emissions and for conserving greenhouse gas reservoirs and sinks. Article 4.2(b) stipulates that national reduction targets may also be achieved in cooperation with other parties and that criteria for this cooperation were to be established at the first conference of the parties (in Berlin, in March 1995). The countries which introduced the JI concept to the negotiations made it clear from the outset that JI should be possible for all parties and specifically mentioned countries with no quantitative targets (Hanisch, 1991, pp.3, 5).

Negotiating Criteria

Since the Framework Convention was passed, criteria and ground rules for JI were repeatedly negotiated in the INC – whose sessions were also open to representatives of non-signatory states – in preparation for the first conference of the parties. At the INC's eighth session in August 1993, for example, 59 countries made a statement on the subject (UN, 1993, p.14), while 31 did so at the ninth session in February 1994 (UN, 1994, p.15).

Resistance from opponents of joint implementation

During the negotiations, it became clear that there was opposition to JI based partly on non-economic, ethical and moral arguments. This opposition was voiced mainly by representatives of many developing countries and NGOs (Encarnacion, 1993; Sharma, 1993; Sanhueza *et al.*, 1994, pp.8f) who saw JI as an attempt by the industrialized countries to buy their way out of reduction commitments and who therefore demanded that JI be restricted to the industrialized countries. This is clearly expressed in the statements made, among others, by Algeria (INC, 1993, pp.4f); Columbia, which spoke for G77 (ibid., pp.16f); Malaysia (ibid., pp.45f); and Nauru (ibid., pp.49ff). There was, however, no united rejection front right from the outset. Mexico was relatively open-minded towards the concept and prepared to include the developing countries as long as a certain proportion of the reduction target was achieved within the industrialized countries and reductions from JI projects were only partly credited (ibid., pp.47f). The Brazilian environment minister Goldemberg was in favour of JI in cases where there were reduction targets which went beyond stabilization targets (McDonald, 1994, p.7). Costa Rica declared itself available for JI projects as early as mid-1994 without reservations (Orlebar, 1994), and concluded the first general agreement on JI with the USA on 30 September 1994 (USA, 1994).

Most environmental NGOs were of the opinion that, if it was not possible to remove JI from the Convention altogether, it would have to be practically prevented by the strictest possible restrictions and by long transition periods. This was evident from a declaration signed by more than 50 European environmental organizations (Climate Network Europe, 1994). Many critics were of the opinion that, by using JI projects to realize low-cost greenhouse gas reduction potentials in developing countries, the industrialized countries could avoid cost-intensive investments at home, thus allowing them to maintain their unacceptable life style.

Some countries maintained that the wording in the Convention on JI could only refer to the industrialized countries from the start, since these were the only countries to have adopted quantitative emission targets. This position was also supported by a legal interpretation of the Framework Convention on

Climate Change by the legal adviser to Samoa (Yamin, 1993, pp.14ff). Yamin stated that the Framework Convention on Climate Change distinction between hard commitments for the industrialized countries and soft commitments for the developing countries[11] was diluted by JI, in that JI entailed an information commitment similar for all countries which is not anchored in the Convention (ibid., pp.4ff). In the critics' opinion, JI could reduce the incentive for structural change in the industrialized countries (Akumu, 1993), thus resulting in a slower rate of emission-reducing innovations (Dubash, 1994, pp.40ff).[12]

Some developing countries were also afraid that JI projects could run counter to their own development priorities. Reforestation projects, for example, would be rejected since they do not involve technology transfer and lead to disputes over land use, if not to 'neocolonialism' (Parikh, 1994, p.5). Also demands arose to limit the crediting of emission reductions achieved by JI projects to the targets of those countries whose development aid is greater than the UN target value of 0.7 per cent of GNP.

It was also argued that power differences between the project participants lead to an inefficient allocation (Loske and Oberthür, 1994, p.7). Dubash took this argument further, concluding that the 'price' of reduction paid to the project participant in the host country would be lowered by the investors' power regarding the level of marginal costs. All the profits were thus appropriated by the investors (Dubash, 1994, pp.36f). Furthermore, some developing countries perceived a danger of JI projects exhausting the 'cheap' reduction options, so that when emission targets are established for their countries at a later date, these targets can only be reached at higher cost.

Many critics wanted to restrict the circle of countries in which JI projects may be carried out to those industrialized countries which have accepted reduction targets and which draw up baseline scenarios.[13] They were equally opposed to JI with countries in which emission levels are falling as the result of a recession (Greenpeace, 1994, pp.2f). This would currently be the case in the countries in transition. Without these restrictions, it is argued, there would be a risk of an increase in global emissions compared to the situation without JI. The host country has an incentive to exaggerate emission reductions, that is, to draw up an unrealistic baseline scenario, and it would thus be impossible to verify the reduction in a country with no emissions target.

---

11. In contrast to the developing countries, the industrialized countries have undertaken a commitment to achieve quantified emission targets and to report at regular intervals on their climate policies.
12. Dubash argues on the basis of the assumption that the investors reduce transfers per emission unit to the level of marginal costs. Moreover, he assumes a proportional relationship between emission reduction costs and the rate of innovation. These assumptions are dubious.
13. Baseline scenarios quantify future emission developments without climate policy measures. They serve as a point of reference for the calculation of emission reductions.

In rejecting JI projects, some developing countries and NGOs implicitly assumed a reference situation which entails extensive finance and technology transfers on the part of the industrialized countries for reduction projects in developing countries (ibid., p.3).

Proposals for compromise

The following proposals for compromise were advanced during the negotiations to preserve at least some of the advantages of JI should there be a political blockade of a far-reaching JI concept. However, all these special provisions reduced the incentives for JI projects.

1. An investing country could have those emission reductions achieved through JI projects credited to a special account, but not to its national reduction target. High credit levels in the special JI account could then be taken into consideration when setting new emission targets for the investing country. Credits towards the national reduction target would thus have taken place implicitly and at a later date.
2. A simultaneous establishment of national and JI targets (dual commitment) was requested by the Netherlands at the eighth INC session (INC, 1993, p.67). This approach would have taken into account the reservations of those countries which demand that the industrialized countries implement emission reduction measures at home first. Moreover, it would have avoided a discussion on credit procedures (Vellinga and Heintz, 1994, p.8).
3. Furthermore, there were demands for the establishment of minimum and maximum quotas for JI projects in relation to national reduction targets. This position was endorsed by a majority of the German Parliament's Enquête Commission on Protection of the Earth's Atmosphere (Enquête-Kommission, 1994, p.12).
4. An incentive for the simultaneous implementation of both domestic and JI reduction measures would be created by crediting the emission reductions achieved through JI projects to the investing country only if a specified domestic target is met at the same time (Moomaw, 1994, p.12). In this case, previous reductions achieved through JI projects would be credited to the investing country if the domestic target were met at a later date.

In view of the massive opposition of many developing countries, the industrialized countries began making concessionary moves at the eighth INC session. While Australia, Canada, Japan, Norway and the USA argued in favour of an immediate start of JI in all the signatory states and direct crediting of reductions towards the stabilisation target (INC, 1993, pp.6f,

12ff, 40ff, 69ff, 100ff), the EU states were considerably more restrained. Finland took up an intermediate position in that it demanded direct crediting but wanted a restriction of JI to the industrialized countries (ibid., pp.23f). The EU statement demanded minimum quotas for the fulfilment of future domestic reduction targets, the exclusion of measures for sequestering greenhouse gases (such as reforestation) and no crediting of reductions until the stabilization target is reached (ibid., pp.9ff). Denmark went one step further by demanding the fulfilment of the development aid target (the provision of development aid to a value of 0.7 per cent of domestic GNP) as a precondition for the crediting of JI projects (ibid., p.21). Together with the Netherlands, Denmark also demanded a pilot phase for the mechanism. France wanted to limit the crediting of JI reductions to the period up to 2000. Furthermore, France demanded quotas for domestic reductions as well as partial crediting and required that all low-cost domestic reduction options be exhausted before JI is permitted (ibid., pp.25ff). Great Britain, Switzerland and Austria demanded its limitation to the industrialized countries (ibid., pp.94, 97ff; Add. 2, pp.2ff). The climate secretariat drew up a compromise solution for the ninth INC session which is reproduced in the Appendix of the present volume.

The ninth session brought a certain convergence of opinion. All the countries which made a statement demanded the introduction of a pilot phase. Some countries, such as the Netherlands, Sweden, the USA and Norway expressed their willingness to push ahead with bilateral JI projects in order to gain concrete experience (INC, 1994, pp.61f, 70, 71ff; Add.1, pp.11f). Within the framework of its action plan for climate change, published by the Clinton administration in October 1993, the USA envisaged an initiative for JI (US Initiative on Joint Implementation, USIJI). The action plan was based on voluntary measures and existing programmes, but contained no tax incentives. Its catalogue of criteria can be found in the Appendix. It follows from the protocol of a US Senate Environment Committee hearing that regulations in individual US states are the only substantial incentive for JI projects (US Congress, 1994, pp.34f). The government saw JI mainly as a trade policy instrument.[14]

The first academic conference on JI, which took place in June 1994 in Groningen (Netherlands) with broad participation of the developing countries and the countries in transition, produced the following results. JI should be voluntary and should respect the sovereignty of the participating countries.

14. 'We agree ... that this [Joint Implementation] is an area of huge opportunity not only for our trade, not only for greenhouse gas emissions, but also for bringing some of our trading partners forward in some of the other interest areas of the United States ... which is to say that we are more likely to have more jobs in America using American technology rather than Japanese or German technology in China' (US Congress, 1994, p.37).

Experience of current JI projects shows that there is a considerable potential for reaching reduction targets set out in the Framework Convention on Climate Change. Given the urgency of climate protection measures, both domestic reduction efforts and JI should be pushed forward. The conference of the parties is called upon to develop objective and checkable criteria for JI projects as well as baseline scenarios and a crediting system. Until this happens, pilot projects are to be supported in order to gain experience and to convince hitherto sceptical countries of the potential of JI projects. Project design should take particular account of developing countries' priorities (Groningen Conerence on Joint Implementation, 1994).

An updated catalogue of criteria was discussed at the tenth INC session in September 1994 (see Appendix in the present volume). Even the supporters of JI proposed a pilot phase which was to last until 1997 or 2000. During this pilot phase, the reductions achieved through JI projects are not to be credited to national stabilization targets. When national targets were redefined following a successful conclusion of the pilot phase, reductions resulting from JI projects could be credited (Cutajar, 1994, p.3). At the eleventh INC session, in February 1995, a split in the ranks of G77 became apparent when several Latin American countries openly called for JI. Only the clear pressure of China brought them back into the common opposing G77 position that called for joint activities without crediting.

THE BERLIN MANDATE AND THE AIJ PILOT PHASE

Originally, it had been hoped that at the Berlin Conference of the parties in 1995 the loose regime of the Convention could be tightened. When it became clear, however, that none of the major players was willing to commit itself, negotiations focused on defining a path to tighter commitments. Against the resistance of the USA and its followers (the so-called 'JUSCANZ' group) the EU pressed for a mandate to negotiate a protocol with binding targets. It won the day when key developing countries, especially India, rallied to its cause and created the 'Green group'. Thus the Berlin Mandate was agreed to negotiate the protocol until 1997.

Concerning JI, a compromise was found: 'activities implemented jointly' (AIJ). A four-year pilot phase without crediting was set up under relatively open criteria (see Appendix).

Sluggish Start and Uneven Results from the AIJ Pilot Phase

Those who thought that, after the Berlin decision, huge funds would flow into AIJ were to be disappointed. While many countries set up national AIJ programmes, and many projects were proposed, only a few have actually been implemented (see Table 1.4). This is so for a variety of reasons: only a few countries provided real incentives, so companies restricted their activities to projects that had a high media interest, opened markets or were profitable on their own.

Before Kyoto there was no fully-fledged incentive system in place (see Table 1.2). The most differentiated system was to be found in the Netherlands. While the USA, Australia, Canada, Japan, Germany, Sweden and Switzerland participate in the pilot phase, they apply no direct incentives apart from covering some transaction costs. The US case is especially interesting as all large AIJ projects have been fostered by strong incentives at the state level, while the federal level has remained inactive. After Kyoto, several countries have put new incentives in place.

Table 1.2 Existing incentives in 1997 and changes in 1998

Country	Tax	Subsidy	Regulation	Voluntary agreement	Subsidy for project preparation
Australia	(-)	*	(-)	X	*
Canada	(-)	(-)	(-)	X	*
Germany	(*)	(-)	(-)	X	X
Japan	(-)	*	(-)	X	X
Netherlands	X	X	(-)	X	X
Norway	(-)	X	(-)	(-)	(-)
Sweden	(-)	X	(-)	(-)	(-)
Switzerland	(-)	X	(-)	(-)	(-)
USA	(-)	(-)	X (state level)	X	X

Note: X = 1997; * = 1998.

The Netherlands grants about US$43 million in direct subsidies for the period 1996–1999, of which US$25 million are allocated to projects in developing countries and US$18 million to countries in transition (Vos, 1997).

Companies are also allowed to depreciate flexibly equipment used for AIJ projects.

Norway is currently subsidizing four AIJ projects with US$7.1 million out of its 'climate fund', and co-finances the World Bank AIJ programme. Switzerland and Sweden also offer subsidies (SWAPP, 1997). None of these subsidy programmes is directed towards private companies, though, and there is no tendering.

Many national pilot phase programmes cover transaction costs such as spreading information, project development and negotiation. This is the prevailing kind of incentive at the moment. For example, Japan finances project preparation with US$0.18 million each. USIJI grants US$25–50 000 per proposal (US DOE, 1997). Transaction cost subsidies often cannot be quantified, but in cases of a large AIJ secretariat, such as USIJI with a staff of 16, they can be substantial. Such indirect subsidies are not only granted by donor countries but also by host countries: the Costa Rican Office for Joint Implementation has a staff of seven (LeBlanc, 1997).

It is likely that the Scandinavian states which apply carbon taxes to industry will allow AIJ-related tax offsets once they have fully defined their participation in the pilot phase. Norway, where 60 per cent of all CO_2 emissions are subject to carbon taxation, ranging from US$16 to US$51/t C (ton of carbon), should be the first to introduce tax incentives, especially as it has been a major driving force in the AIJ debate. Nevertheless, tax concessions are not yet feasible, as the Ministry of Finance is strongly opposing them (Lunde, 1997). Swedish and Finnish industry could have incentives of around US$36/t C which would imply a huge demand for AIJ. So far, no plans for tax concessions have surfaced.

In the Netherlands, income tax exemption for income from AIJ projects has been granted. This will be a very powerful incentive for 'no-regret' projects. Astonishingly, crediting against the energy tax is not envisaged.

In several states of the USA, 'externality adders' are a powerful incentive for electric utilities to invest in carbon offsets including AIJ. A total of 13 states have established explicit values for the evaluation of externalities, some of which vary quite considerably (Table 1.3). In the majority of states, flat-rate externality adders expressed in percentage terms or in US$/kWh are used when approving new power stations. Five states have quantified externalities for individual substances, although not all greenhouse gases are covered.

Table 1.3 Evaluated externalities of CO_2, CH_4, N_2O and NO_x emissions (US$/t)

State	Authority	CO_2	CH_4	N_2O	NO_x	Prices
California	Public Utility Commission	9	0	0	7 495–31 448	1992
	Energy Commission	7.6	0	0	730–16 076	
Massachusetts		22	220	3 960	6 500	1989
Nevada		22	220	4 140	6 800	1990
New York		8.6	0	n.a.	6 524	1992
Oregon	Bonneville Power	6	0	0	69–884	n.a.
Wisconsin		17	165	2 977	0	1992

Source: New York State Energy Office (1994, p.153).

Externality adders have led to the biggest private AIJ projects to date with an aggregate investment of more than US$6 million. The first utility which invested in AIJ did it because it feared an externality adder long before the term 'AIJ' was coined – in 1988! Then, AES Corporation invested US$2 million in afforestation in Guatemala to offset its emissions of a 183MW coal-fired power plant in Connecticut. It later invested the same amount in Paraguay. Tenaska invested US$1 million in one of the few funded USIJI projects – the ECOLAND project in Costa Rica (Trexler and Associates, 1994; LeBlanc, 1997).

The first offset requirement was introduced in 1995 in New Zealand, where emission from a 400MW gas-fired plant will be compensated through (domestic) afforestation (Brasell, 1996).[15] Oregon had a fierce competition between three offset portfolios in 1996. The winning proposal includes an AIJ component – rural photovoltaics in developing countries (Trexler and Associates, 1996). It recently put offset legislation in place. The Ohio Public Utilities Commission has allowed Cincinnati Gas&Electric to defer costs of their commitment to the Rio Bravo AIJ project to its customers (*PUCO News*, 1997).

The current deregulation in US electricity markets will reduce regulation of the type described. As deregulation is also done in other parts of the world,

15. It only applies, though, if aggregate emissions from the electricity sector rise above the level reached before the plant started operation.

the possibility of using its potential for AIJ incentives will be curtailed. Thus the only strong incentives currently existing are likely to vanish soon.

The Netherlands registers emission reduction from AIJ projects and certifies them (Vos, 1997). If international crediting is allowed in the future, these certified credits could become valuable. So far, no certification has taken place. When the voluntary agreements with industry are renegotiated in 2000, AIJ could be included. Costa Rica has already started to certify emission reductions from three large, sector-wide AIJ projects to attract different investors which could be interested in 'creditable, tradable offsets (CTOs)' (LeBlanc, 1997). This allows different investors to invest in the same project. In Germany, the Federation of German Industry includes AIJ in its voluntary agreements but has not yet clarified certification and baseline issues. The same applies to voluntary commitments in Australia, Canada, the Netherlands and the USA.

Table 1.4 The AIJ pilot phase over time

	1995	1996	1997	1998	1999*
Accepted projects	10	16	61	95	122
Projects actually being implemented**	0	3	13	60	86
Investing countries	3	3	5	8	11
Host countries	7	7	12	24	34
Countries in transition	5	5	7	10	11
Share of countries in transition of all projects (%)	60	50	74	72	65
Planned emission reduction (million tons CO_2)+	23	111	140	162	217
Share in countries in transition (%)	56.5	39.5	32.6	31.3	24.3

Notes:
* 22 July.
** These are estimates as no reliable information exists. The implemented projects tend to be small projects in countries in transition.
+ The emission reduction actually implemented is much lower (see previous note).

Sources: Own calculations from data in UNFCCC (1997a, 1998, 1999) and personal communication from Kai-Uwe Schmidt (22 July 1999), Anon. (1995) and Project Carfix (1995).

Besides lack of incentives, transaction costs were very high in the beginning. Moreover, many attractive host countries did not give their approval to projects. In the time between Berlin and Kyoto the opposition of major developing countries such as India and China against JI with crediting of emissions reduction hardened again many analysts thought that JI would – if at all – only be possible between countries with emission targets. Thus, the interest for AIJ concentrated on countries in transition.

UNCTAD AND EMISSION TRADING

During the whole debate on joint implementation, emission trading was only discussed in academic circles – and even this debate was not very intense. The only exception was a bold initiative by the United Nations Conference on Trade and Development (UNCTAD), an outsider in the climate policy debate. UNCTAD had already published a report on the advantages of emission trading in 1992 (UNCTAD, 1992) and distributed it at the Rio Conference, where it had not got much attention. It did not give up, though, and published a follow-up with a very detailed concept of rules in 1994 (UNCTAD, 1994) as input for Berlin. Although trading still did not get much attention at that time, UNCTAD went on and published a fully elaborated set of draft rules at the second conference of the parties in Geneva in 1996 (UNCTAD, 1996). By then the USA had begun to stress trading as a flexibility instrument and thus the UNCTAD ideas gathered force. In the run-up to Kyoto, UNCTAD founded an Emissions Trading Forum, together with the NGO Earth Council, and convened several meetings of stakeholders in 1997. It thereby paved the way for a better understanding of the issues surrounding trading.

THE KYOTO PROTOCOL: A WHOLE MENU OF FLEXIBILITY INSTRUMENTS

In the course of the negotiations on the Berlin Mandate, the USA made a strong drive for inclusion of flexibility measures in the protocol, while only proposing a weak stabilization target for 2010 late in 1997. In contrast, the EU proposed a 15 per cent reduction for all industrialized countries by 2010, with an intermediate target of 7.5 per cent by 2005. The overall reduction target of 5.2 per cent for 2010 achieved at Kyoto would not have been reached without the strong EU position. Nevertheless, the EU backed down on its opposition to differentiation of targets and flexibility instruments. It only stuck to its demand that a 'concrete ceiling' should limit the use of the flexibility instruments. Therefore the Kyoto Protocol contains all the instruments that were proposed by the USA. Four distinct possibilities have been allowed by the Kyoto Protocol (UNFCCC, 1997c): bubbles, emission trade, joint implementation and the clean development mechanism.

Bubbles

The first and most far-reaching is an agreement on joint targets, or 'bubbles' (Art. 4). Countries that want to set a common target can set up a bubble and have to declare it at the latest when they ratify the Protocol. The bubble remains binding throughout the commitment period. So far, the EU agreement is the only bubble. Current negotiations between Russia, on the one hand, and the USA, Canada, Japan and New Zealand, on the other, could lead to the creation of additional bubbles. This could lead to a reduction of the potential for international emission trade. If the target is not reached by the bubble, each state individually is responsible for the target allocated by the bubble agreement. In the case of the EU, which also is a party to the Convention, binding sanctions against member states that fail to reach their target would be necessary.

Emission Trade

A fairly elaborate article on emission trade was thrown out late in the process (the 'lost Article': see Appendix) owing to the resistance of developing countries. The USA only managed to get a very crude and open formulation in at the last minute that states that emission trade is only open to industrial countries (Art. 17). Article 17 stipulated that 'emission reduction units' (that is, permits) are to be added to the budget of the buyer and to be deducted from the budget of the seller. Trade has to be based on common methods for inventories using Intergovernmental Panel on Climate Change (IPCC) methodology and verification to prevent trade with fictitious emission reduction. Sanctions have to be used to avoid violation of methodology.

Joint Implementation

Article 6 allows industrial countries to acquire emission permits through investment in projects in other industrial countries. The criteria for projects are the same as in the AIJ pilot phase. Emission permits created in this way are to be considered equal to emission permits from emission trade under Article 17. Emission permits are not created if annual reporting requirements have not been met or the reports do not comply with the binding rules. If a review team has doubts about additionality of emission permits they shall still be tradable but are frozen until the doubts are resolved.

Clean Development Mechanism

The clean development mechanism (CDM) has been rightly called the 'Kyoto surprise' (Estrada-Oyuela, 1998). While the developing countries had rejected any notion of JI with credits until Kyoto, two strands of negotiation joined there which led to their de facto acceptance of the concept. On the one hand, Brazil had tabled a proposal in early 1997 that envisaged penalties for industrial countries if they did not reach the proposed, strict emission targets. The penalties were to be channelled in a 'Clean Development Fund' which would finance adaptation measures in countries which would be most seriously affected by climate change. On the other hand, the USA remained firm in its wish for JI with credits with developing countries. With the aid of Costa Rica, it managed to convince the Brazilians to change their proposal to a clean development mechanism that would still finance adaptation but act as a vehicle for JI with crediting, as the penalty approach would not be possible.

Article 12 of the Kyoto Protocol outlines the CDM. It states in paragraph 3 that investing countries get credit for certified emission reductions from CDM projects provided 'benefits' accrue to the host country (Art. 12 (3a)). Crediting will only be allowed until a certain percentage of the emission target is reached (Art. 12 (3b)) that remains to be defined. It is unclear whether crediting up to this quota is in full or only partial. Besides countries, companies are allowed to invest and execute projects (Art. 12 (9)). The CDM will cover its administrative budget through project revenues. Moreover, a 'part' of these revenues will be used 'to assist developing country Parties that are particularly vulnerable to the adverse effects of climate change to meet the costs of adaptation' (Art. 12 (8)). It remains open who does certification of emission reduction, but verification will be done by independent bodies (Art. 12 (7)). The project criteria remain the same as for AIJ (Art. 12 (5)).

There are two general options for CDM: bilateral and multilateral. The bilateral option allows countries to negotiate a framework agreement setting criteria and rules for crediting (see Figure 1.2). Projects are negotiated freely between entities of both countries. In the multilateral option, investing countries make contributions to an independent fund (see Figure 1.3). Other countries can now offer projects and so compete for the fund's resources. Projects are selected according to their emission reduction efficiency, with positive externalities being taken into account in the case of equally efficient projects. For the duration of the project, each investor country receives a credit proportional to its share of the project portfolio. Project risks would also be pooled, with the investor countries being required to pay a corresponding insurance surcharge. The necessary verification could be carried out multilaterally or by private auditors (Mintzer, 1994, p.46 described under the term 'mutual fund').

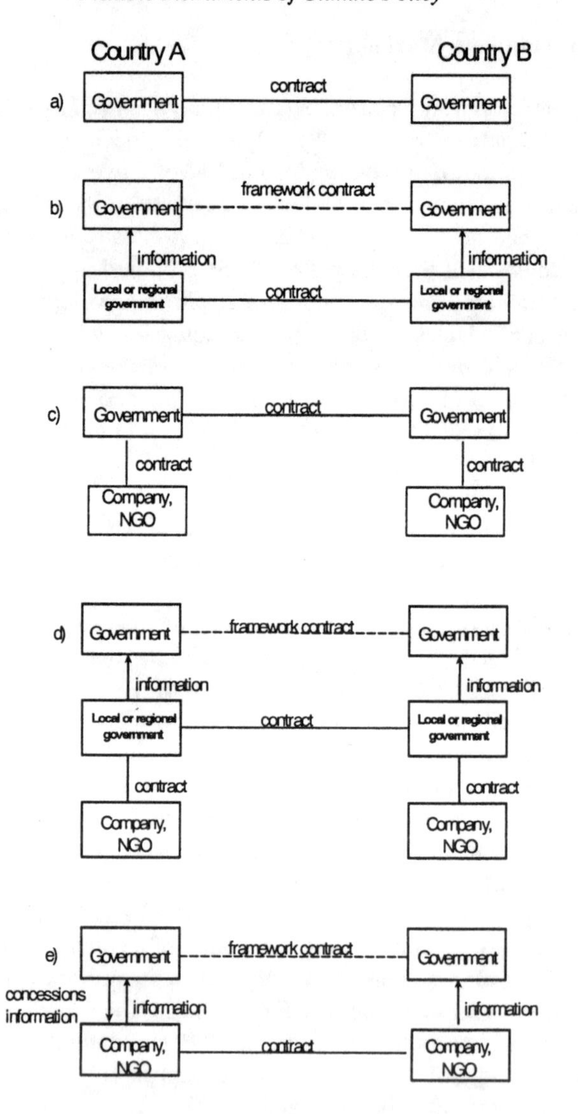

Figure 1.2 Bilateral CDM: forms of contractual agreement

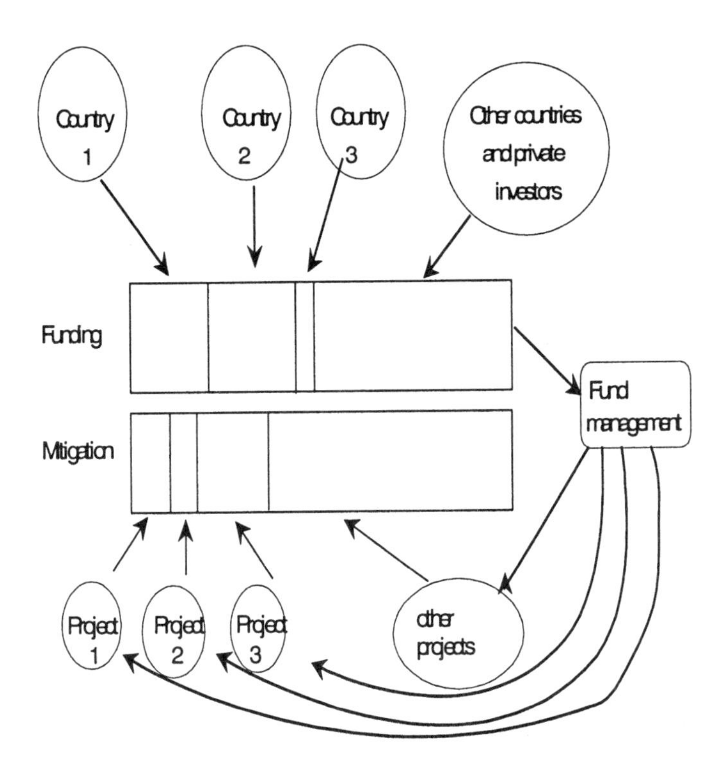

Figure 1.3 *Multilateral CDM*

BUENOS AIRES: A WORKPLAN FOR THE TURN OF THE CENTURY

Many stakeholders in the climate negotiations had hoped that at least some of the open issues in the definition of the flexibility instruments could be solved at the 1998 conference of the parties at Buenos Aires. This hope was owing to the flurry of activity concerning the interpretation of the flexibility instruments. There were dozens of specialized workshops, and they were also discussed in a flood of publications between February and October 1998 (see, for example, Tietenberg *et al.*, 1999; Dutschke and Michaelowa, 1998a, 1998b, 1998c; CIEL/Euronatura, 1998; Greenpeace, 1998; EDF, 1998; Matsuo, 1998). This led to a better understanding of the possibilities that flexible instruments can achieve but also sharpened the eyes for possible loopholes owing to flawed design or misuse of the instruments.

The UNCTAD initiative gained momentum with a packed session of the Greenhouse Gas Trading Forum in May and the creation of a new International Working Group on the CDM. At Buenos Aires, it set up the International Emissions Trading Association, a step that drew criticism from the USA-based Emissions Marketing Association which feared monopolization of the market. Moreover, the United Nations Industrial Development Organization (UNIDO) and UNEP (UN Environmental Programme) joined the UNCTAD initiative. Many other stakeholders that had been reluctant until now entered the debate, mainly from the commercial sector. Several big companies announced emission targets and intra-company emission trade. The pioneer in this respect was British Petroleum, but Shell followed suit.

The Buenos Aires conference was paralysed, however, by the conflict between the USA and developing countries over emission targets for the latter. Owing to the insistence of many interest groups in the USA, US ratification of the Protocol is unlikely without targets for developing countries. The domestic conflict thus spilled over to the international level. The developing countries were not interested in moving the flexibility instruments forward unless the USA would drop the target issue. Only in the case of the CDM did a group of Latin American and African countries press for an 'interim phase' with crediting. The EU tried to bridge the gap between the USA and the developing countries, but still stuck to its demand for a concrete ceiling. In the end, it could only be decided that decisions on the open issues will be taken by 2000 and that the CDM should have priority. Thus little time remains in which the debate can be further refined.

PROJECT TYPES FOR CDM AND JI

The wide range of meaningful projects which can lead to a creditable reduction of greenhouse gas emissions is outlined below. Besides emission reduction measures, sequestration measures could be taken into account. For the time being, it is not clear whether the latter are allowed under the CDM, as many developing countries and environmental NGOs oppose them. The IPCC will publish a special report in 2000 that discusses the suitability of land use and forestry as measures of climate policy.

Emission Reduction Measures

Efficiency improvements in energy generation based on fossil fuels

These include such measures as increasing the efficiency of power stations and steam generators. For example, while the average efficiency of German coal-fired power stations is approximately 35 per cent, the figure in China is only approximately 20 per cent (Michaelis, 1992, p.510). Implementation of state-of-the-art power generation technology with an efficiency level of between approximately 42 per cent (lignite power station) and 55 per cent (gas-fired combined heat and power station) can thus achieve an emission reduction of over 50 per cent (Matthes, 1994, p.51). In the case of combined heat and power generation, efficiency levels can be increased to over 80 per cent (OECD, 1994, p.98). By transferring know-how and expertise, it is possible to ensure that technically feasible efficiency levels are also achieved in practice. Similarly, emissions caused by the production and transport of fossil fuels can also be reduced (see Table 1.5). Since, for instance, avoidable losses of 2–4.5 per cent occur during the extraction and transport of natural gas in the CIS countries, Central Asia has great potential for CDM projects (Lelieveld and Crutzen, 1994, p.438).

Table 1.5 Emission factors for CO_2 (in % of emissions on combustion)

Fuel	Extraction	Processing	Transport	Total
Crude oil	0.6	10.9	3.5	15.0
Hard coal	2.1	0.3	2.1	4.5
Natural gas	10.9	5.8	4.6	21.3

Source: New York State Energy Office (1994, p.558).

Conversion of energy generation to fuels which emit fewer greenhouse gases or none at all

Conversion from coal to gas or to nuclear energy is an example. In the long-term, regenerative energies such as sun, wind, water and biomass have a great potential for reducing greenhouse gases. According to International Energy Agency projections, renewable energies account for 13 per cent of the world's energy use in the year 2020 business-as-usual case (IEA, 1998, p.27). As the Indian example shows, with over 1000MW of wind power installed by the end of 1998, this potential can be tapped quickly if an appropriate political framework exists. Demands for emissions to be taken into account during the entire project cycle – for example the construction of a power plant – when calculating greenhouse gas reductions discriminate against renewable energies. Owing to the dominance of fossil fuels in the world's energy systems it has, for example, so far been impossible to use building materials whose manufacture caused no greenhouse gas emissions.

Reduction of energy consumption by means of more efficient manufacturing processes

Process innovations which accompany the manufacture of new products lend an inherent momentum to improvements in energy efficiency. Moreover, given certain incentives this momentum can be accelerated still further. This is clearly demonstrated by the pioneering project 'Ökoprofit Graz' (Eco-Profit Graz). The Austrian town of Graz, in cooperation with the local Institute of Science and Technology, is carrying out a detailed analysis of production processes in local companies, at the same time identifying efficient, cost-saving reduction options. Of 54 options, 30 were implemented immediately. Participating companies receive an award which is beneficial to their public image. Although the main aim of the project has so far been to reduce waste levels, the approach itself can easily be applied to reducing energy consumption (Niederl and Schnitzer, 1994). The cement industry in particular has a high efficiency and emission reduction potential, as it has the highest share of energy-related carbon emissions in many developing countries.

Changing consumption behaviour in favour of goods which require low levels of energy during production and in use[16]

This necessitates permanent behavioural changes. Private motorized transport, for example, could be reduced by providing efficient local public transport. In developing countries, electrical transmission losses can be reduced relatively easily, and demand for biomass (as a cooking fuel) from non-sustainable sources can be reduced by using more efficient cookers. Demand-side activities, however, must be supported by the removal of energy subsidies which have hitherto led to artificially low energy prices in many countries. Furthermore, efficient state regulatory structures are helpful for providing incentives to conserve energy.

Changes in agriculture

Use of appropriate feed, for example, can drastically reduce methane emissions, which are particularly high in intensive cattle farming. Methane production in wet rice farming depends largely on the kind of irrigation used. Also, N_2O production is largely determined by the use of agricultural fertilizer. Large differences in marginal reduction costs – and thus great potentials for joint implementation projects – can be expected in the

16. The same, of course, applies to services.

agricultural sector in particular, where the protection of national markets leads to high levels of inefficiency.

Sequestration Measures

Changes in land use aimed at sequestering higher levels of carbon

An example of this is reforestation. Forest soil contains approximately 25 per cent more carbon than the same kind of soil in agricultural use. This corresponds roughly to the amount of carbon sequestered by the trees growing there (Houghton and Woodwell, 1994, p.35).

Preservation of natural greenhouse gas sinks under immediate threat

These sinks include swamps and marshland, forests, fields and meadows. However, questions of the permanence of sequestration activities and of the baseline scenario are more difficult to solve than in the case of reduction measures. It is relatively difficult to estimate how much gas is sequestered, since rates of CO_2 sequestration differ widely according to particular types of vegetation or soil. Estimates have nonetheless become increasingly accurate in the last few years. Given the long observation time-spans involved, the risk of a sink being destroyed completely is probably greater than the risk of no abatement being achieved in a reduction project. Allowances must be made for this risk in the form of insurance cover which extends beyond the duration of the mere project itself (TransAlta Corporation, 1994, p.37). On the other hand, sequestration activities are indubitably the most cost-efficient at present. The very first CDM project of all, which was initiated in Guatemala in 1988 even before the term 'JI' had been introduced, is a reforestation project. The US energy utility Applied Energy Services had a local NGO plant 52 million trees in compensation for a new coal-fired power station.[17] It recognized, even then, that the social environment has to be respected and so financed development programmes for the resident population. Primary forests were also bought up (Moomaw, 1994, p.11).

Monocultural reforestation – which may at first appear the most cost-efficient – is inadvisable from an ecological point of view, even if it can reduce the pressure on primary forests. Monocultures can lower the water table (for example, in the case of eucalyptus) and are of little benefit to the local population since they allow no agricultural usage (Carrere, 1993). Projects which enable a combination of forestry and agriculture, known as agroforestry, would be ideal in this context, although they would probably be considerably more expensive than simple reforestation. In the case of private enterprise sequestration projects, state grants should therefore help make

17. It is interesting to note that 20 per cent overcompensation was the target. The introduction of an externality adder provided the incentive.

diversified reforestation projects more attractive. This sort of procedure is beneficial for the economy as a whole since more diversified projects involve greater positive externalities. The preservation of threatened areas of primary forest can bring even greater positive externalities. However, the possibility of countries clearing primary forests simply to gain land areas for reforestation must be ruled out (Loske and Oberthür, 1994, p.10; Bohm, 1994, p.191). A realistic baseline scenario is extremely important in this context in order to estimate the emissions caused by alternative land use following the deforestation which would have taken place had there not been a project. Also to be avoided is a situation whereby countries propose rainforest protection as a CDM project while simply moving their previous deforestation activities to other untouched areas of forest (Goldberg, 1994, p.3). The quality of verification is of crucial importance when it comes to the preservation and enlargement of greenhouse gas sinks.

In the AIJ pilot phase the projects can be classified as in Table 1.6.

Table 1.6 Share of project types of AIJ and their share in overall emissions reduction (per cent)

	1995	1996	1997	1998	1999*
Forest protection and reforestation	30 (84.5)	25.0 (68.4)	13.4 (58.5)	11.6 (52.2)	9.8 (64.9)
Afforestation	10 (1.3)	6 (0.2)	1.6 (0.2)	1.1 (0.2)	1.6 (0.1)
Agriculture	0	0	1.6 (<0.1)	2.1 (1.8)	1.6 (1.4)
Fuel substitution	20 (10.7)	12.5 (2.2)	3.2 (1.7)	3.2 (1.8)	5.7 (1.7)
Methane capture	0	6.3 (27.0)	3.2 (21.3)	2.1 (18.6)	3.3 (14.4)
Energy efficiency	30 (2.5)	25 (1.7)	34.4 (2.6)	37.9 (4.8)	40.2 (3.5)
Renewable energy	10 (1.0)	25 (0.5)	42.6 (15.6)	42.1 (20.5)	37.7 (13.9)

Note: *22 July.

Sources: Own calculations from UNFCCC (1997a, 1997b, 1998, 1999) and personal communication from Kai-Uwe Schmidt (22 July 1999), Anon. (1995) and Project Carfix (1995).

POSSIBLE INVESTORS IN CDM PROJECTS

Companies

The private sector of the industrial countries is the natural investor in CDM provided incentives are in place. Companies will invest if the net cost per ton reduced (including transaction cost) is lower than the tax or regulatory benefit. They will try to maximize side benefits such as opening markets or building synergies with existing operations.

Municipalities

The local area offers promising opportunities for public CDM projects. Local administrations are often large-scale emitters of greenhouse gases because they own their own energy utilities and administer a great number of public buildings. Communal enterprises with high levels of energy consumption are active in many sectors. Local public transport firms in a number of countries, for example, are run by local authorities. Furthermore, these authorities can directly influence their inhabitants' greenhouse gas emissions via construction development and traffic plans or local taxes. Many towns and cities have recognized this potential and are making efforts to draw up climate protection concepts. Intensive use is already being made of the possibilities of international collaboration. In 1990, over 140 towns and cities in more than 40 countries joined to form the International Council on Local Environmental Initiatives (ICLEI). In 1992, ICLEI initiated a local CO_2 reduction project with the participation of 14 local communities. As a first step, all the participating communities drew up an energy consumption profile for the reference year 1988. On the basis of these profiles, a catalogue of emission reduction measures was constructed which was then evaluated and used to calculate a reduction target. The project has set itself an aggregate reduction target of 20 per cent by the year 2005. Targets for individual cities go still further. Saarbrücken and Hanover are aiming for 25 per cent, Copenhagen for 30 per cent (Heidelberg Conference, 1994, p.23).

Besides the ICLEI CO_2 reduction project, many communities, especially in the industrialized countries, are in the process of drawing up emission inventories and quantitative emission targets. In doing so, they are laying important foundations for JI and CDM projects, since inventories and targets can be used as baseline scenarios. Another important aspect is that local projects are much easier to monitor than large-scale projects. The ICLEI recognized this fact very early. At a conference of European mayors from 83 towns and cities which took place in Amsterdam in 1993, it was decided, among other things, to enter into energy partnerships with communities in the developing countries and the countries in transition and, within this framework, to transfer know-how and financial resources (ICLEI, 1993a, p.25). Concrete formulation of JI projects was recommended as soon as criteria have been established. However, only large-scale projects involving several towns and cities were said to have chances of success (ICLEI, 1993b, p.29). The Climate Alliance (Klima-Bündnis), which is a further group of 350 European communities which have committed themselves to a CO_2 reduction target of 50 per cent during 1987–2010, emphasizes development collaboration (Klima-Bündnis, 1994, pp.2f). However, it has so far been very critical concerning CDM projects. As a matter of priority, local JI and CDM projects could cover the following areas (OECD, 1994, pp.19f, 43f) which are illustrated with selected examples:

1. Reduction of energy consumption in municipal buildings. For instance, an information campaign aimed at caretakers and janitors or the implementation of the simplest insulation measures can lead to considerable savings (Demczuk, 1994).
2. Combined heat and power generation in combination with district heating systems. In Odense, Denmark, for instance, 95 per cent of dwellings are connected to district heating (OECD, 1994, p.113). Although such a development can only be achieved in the longer term, it is perfectly realistic for 50 per cent of dwellings to be connected in 20 years, as the example of Brescia, Italy demonstrates (ibid., p.106).
3. Renewable energies. In order to support solar energy, the town of Aachen guarantees a cost-covering price for solar energy fed into the power grid. The public utilities in Norrköping, Sweden, are building a biomass power station (ibid., p.160).
4. Know-how transfer and public information. In Copenhagen, the Centre for Energy and Environmental Issues deals with over 1500 enquiries on renewable energies and energy conservation a day (ibid., p.73). In The Hague, unemployed people are being retrained as energy conservation consultants who then inform specifically low-income families about ways of saving energy (ibid., pp.79f).
5. Passive solar energy use or shielding by means of appropriate construction planning. The example of Schiedam, Netherlands, is remarkable: through skilful planning and efficiency standards, passive use of solar energy led to a fourfold increase in energy efficiency in new buildings over a period of 20 years (ibid., pp.35f).
6. Reduction of transport requirements by means of appropriate, integrated transport planning and functional variety within urban districts. By consistently strengthening its bus-oriented public transport, Curitiba, a Brazilian city with over 1.5 million inhabitants, has succeeded in

attracting 70 per cent of its commuters on to the buses. Transport-related energy consumption in Curitiba is 20 per cent lower than in comparable Brazilian cities. Moreover, there is a high level of decentralized employment (ibid., pp.127f).

Financial incentives are necessary in order to utilize the potential of local JI and CDM projects. For enterprises under municipal control, such incentives arise from tax benefits corresponding to those for private companies. Incentives can be provided in other areas by increasing allocations of funds. For example, greenhouse gas reductions from a JI or CDM project could lead to an allocation from the national government corresponding to the tax benefits made to a private company. This allocation can be financed by greenhouse gas tax revenues.

Non-governmental Organizations

In the past decades, the worldwide importance of non-governmental organizations, that is, non-profit-making, voluntary, private sector organizations, has increased immensely. While they have played a decisive role in asserting and institutionalizing environmental goals in the industrialized countries, they ensure at least a minimum of pluralism in the developing world. Demonstrating their interest in global environmental issues, over 1400 NGOs took part in preparations for the UN Conference on Environment and Development. Almost half of them were from developing countries (Stahl, 1994, pp.248f). However, the interests of individual NGOs differ widely. When the GEF (Global Environment Facility) was being restructured, NGOs from developing countries were opposed to greater NGO participation because they feared domination by NGOs from industrialized countries (Ling, 1994, p.5)! Many NGOs in developing countries have a critical attitude towards technologies from industrialized countries because they fear new dependencies, and demand the self-determined development of appropriate technology (Dubash and Oppenheimer, 1992, p.267). Given the great variety of NGOs, it is impossible to form a single general judgement on their efficiency. Some authors are increasingly critical in their observations of particular NGOs (Hanisch and Wegner, 1994).

Regardless of such opinions, NGOs in the industrialized countries are indispensable as important conveyors of funds since they now have access to considerable financial resources. At the end of the 1980s, the development-oriented NGOs alone placed over US$6 billion p.a. at the disposal of their partners in the developing countries (Wegner, 1994, p.119). If we include the environmentally oriented NGOs, a transfer potential of over US$10 billion can be assumed, implying a considerable potential for CDM projects. In addition, NGOs can carry out projects in areas which are unattractive for companies or governments. These can be projects with a large number of participants and small budgets, or projects which run counter to government priorities. Some NGOs are already carrying out de facto CDM projects today. For instance, the World-wide Fund for Nature (WWF) operates technology transfer in India for specific, particularly energy-intensive branches of industry; pilot projects are already beginning in the ceramics, paper and cement industries (Singer, 1994, pp.16f). Various types of energy-saving cookers were distributed in Kenya by the environmental organization KENGO in order to test the practicality of their construction. This experience was used to design the ideal cooker, of which over 0.5 million have since been sold. With fuel savings of 30 per cent, additional costs of 100–200 per cent and twice the life-span of a traditional cooker, the new cooker pays for itself within a few months (Chege, 1993).

NGO-supported projects also seem to be very efficient when it comes to information dissemination and training, since the transfer of know-how to large groups of the population can produce significant long-term effects. This is true for both the industrialized and the developing countries. For instance, NGOs from industrialized countries could collect information on energy-efficient technologies and pass it on to NGO-run regional information centres in developing countries where joint research could also take place (Dubash and Oppenheimer, 1992, p.278). The WWF already supports training centres for the improvement of energy efficiency in Russia, Poland, Bulgaria, the Czech Republic and Ukraine. Similarly, NGOs have a good fundamental basis for achieving a demand-side reduction of greenhouse gas emissions through energy conservation because they enjoy a far more trusting relationship with the population than do state-run entities. The chances of such projects reaching large sections of the population would appear very high.

NGOs should therefore be able to conclude contracts for CDM projects autonomously. It is imperative that they be taken into consideration when it comes to formulating the CDM framework. Their role is especially promising wherever it is possible to bypass inefficient or even corrupt state bureaucracies. Ideally, CDM projects should therefore be arranged in countries directly with NGOs acting as project partners.

As investing NGOs normally pay no taxes because of their non-profit status, they should, on proving an emission reduction, receive a reimbursement in their home country which corresponds to the tax benefits granted to companies as a result of CDM projects. NGOs will therefore have an incentive to carry out CDM projects if the costs of a project are lower than the reimbursement. This will probably be the case quite frequently since NGO labour costs are considerably lower than those of companies or state entities owing to the honorary nature of their work. The NGO can then use any surplus for other projects. In the case of a regulatory solution, NGOs would need to receive a subsidy as an incentive, assuming that such an incentive is politically desirable.

REFERENCES

Akumu, Grace (1993), 'Joint implementation: Mitigation strategy or hidden agenda?', *Eco* (4), 3.

Anonymous (1995), 'Planned and ongoing JI (pilot) projects', *Joint Implementation Quarterly*, 1(2), 14.

Barrett, Scott (1994), 'Climate change policy and international trade', in Akihiro Amano, Brian Fisher, Masahiro Kuroda, Tsuneyuki Morita and Shuzo Nishioka (eds), *Climate Change: Policy Instruments and their Implications*, Proceedings of the Tsukuba Workshop of IPCC Working Group III, Tsukuba, pp.15–33.

Bohm, Peter (1994), 'On the feasibility of joint implementation of carbon emissions reductions', in Akihiro Amano, Brian Fisher; Masahiro Kuroda, Tsuneyuki Morita and Shuzo Nishioka (eds), *Climate Change: Policy Instruments and their Implications*, Proceedings of the Tsukuba Workshop of IPCC Working Group III, Tsukuba, pp.181–98.

Brasell, Robin (1996), 'New Zealand's net carbon dioxide emission stabilisation target', *Agenda*, (3), 329–40.

Bundesverband der Deutschen Industrie (1991), *Initiative der deutschen Wirtschaft für eine weltweite Klimavorsorge*, Cologne.

Carrere, Ricardo (1993), 'The dangers of monoculture tree plantations', *Third World Resurgence*, (40), 4–5.

Center for International Environmental (1998), *Responsibility or Non-compliance under the Kyoto Protocol's Mechanism for Cooperative Implementation*, Washington.

Chege, Nancy (1993), 'Im Brennpunkt: Herde für die Dritte Welt', *World Watch* (12), 44–6.

Climate Network Europe (eds) (1994), *Joint Implementation – from a European NGO perspective*, Brussels.

Cutajar, Michael (1994), 'INC to review national efforts to implement treaty', *UN Climate Change Bulletin* (4), 2–3.

Demczuk, Jerzy (1994), 'Energy bus drives energy conservation in Poland', *E-notes* (1), 3–5.

Dubash, Navroz (1994), 'Commoditizing carbon: Social and environmental implications of trading carbon emissions entitlements', Masters thesis, Berkeley.

Dubash, Navroz and Michael Oppenheimer, (1992), 'Modifying the mandates of existing institutions: NGOs', in Irving Mintzer (ed.), *Confronting Climate Change*, Cambridge, pp.266–79.

Dutschke, Michael and Axel Michaelowa (1998a), 'Interest groups and efficient design of the Clean Development Mechanism under the Kyoto Protocol', HWWA Discussion Paper no. 58, Hamburg.

Dutschke, Michael and Axel Michaelowa (1998b), Creation and sharing of credits through the Clean Development Mechanism under the Kyoto Protocol, HWWA Discussion Paper no. 62, Hamburg.

Dutschke, Michael and Axel Michaelowa (1998c), 'Issues and open questions of greenhouse gas emission trading under the Kyoto Protocol', HWWA Discussion Paper no. 68, Hamburg.

Encarnacion, Rene (1993), 'A diversionary tactic', *Joint Implementation – an Eco special report*, 1.

Enquête-Kommission 'Schutz der Erdatmosphäre' des Deutschen Bundestags (1994), *Empfehlungen für die 1. Vertragsstaatenkonferenz zum Rahmenübereinkommen über Klimaänderungen (Klimarahmenkonvention) vom 28. März bis 7. April 1995 in Berlin*, Bonn.

Environmental Defense Fund (EDF) (1998), *Cooperative Mechanisms under the Kyoto Protocol*, Washington.

Estrada-Oyuela, Raul (1998), 'First approaches and unanswered questions', in José Goldemberg (ed.), *Issues and Options: The Clean Development Mechanism*, New York, pp. 23–30.

Frese, Walter (1994), 'Die Treibhaus-Fenster schließen sich', MPG-Presseinformation, 14 February.

Germanwatch (1994), 'Acht Kriterien für den Klimagipfel Berlin 95', Presseerklärung, 31 May.

Goldberg, Donald (1994), *Joint Implementation: The NGO Perspective*, Washington.

Greenpeace (1994), *The Joint Implementation Pilot Phase: A Critical Approach*, Geneva.

Greenpeace (1998), *Making the Clean Development Mechanism Clean and Green*, Buenos Aires.

Groningen Conference on Joint Implementation (1994), *Groningen Statement on Joint Implementation*, Groningen.

Grubb, Michael (1989), *The greenhouse Effect: Negotiating Targets*, London.

Grubb, Michael and Adam Rose (1992), 'Introduction: nature of the issue and policy implications', in United Nations Conference on Trade and Development (ed.), *Combating Global Warming*, New York, pp.1–10.

Grubb, Michael and James Sebenius, Antonia Magalhaes and Susan Subak (1992), 'Sharing the burden', in Irving Mintzer (ed.), *Confronting Climate Change*, Cambridge, pp.306–21.

Hanisch, Rolf and Rodger Wegner (1994), *Nichtregierungsorganisationen und Entwicklung*, Schriften des Deutschen Übersee-Instituts Hamburg no. 28, Hamburg.

Hanisch, Ted (1991), 'Joint implementation of commitments to curb climate change', CICERO Policy Note 1991(2), Oslo.

Harper, Stephen (1994), 'Tradable permits: practical lessons from the US experience', in Akihiro Amano, Brian Fisher, Masahiro Kuroda, Tsuneyuki Morita and Shuzo Nishioka (eds), *Climate Change: Policy Instruments and their Implications*, Proceedings of the Tsukuba Workshop of IPCC Working Group III, Tsukuba, pp.132–44.

Heidelberg Conference 'How to combat global warming at the local level' (1994), *Compendium of Abstracts*, Heidelberg.

Hoel, Michael (1993), 'Efficient climate policy in the presence of free riders', Memorandum from Dept. of Economics, no. 4, University of Oslo.

Houghton, Richard and George Woodwell (1994), 'Forests as carbon sinks', in Kilaparti Ramakrishna, (ed.), *Criteria for Joint Implementation under the Framework Convention on Climate Change*, Woods Hole Research Center, pp.35–40.

Intergovernmental Negotiating Committee for a Framework Convention on Climate Change (1993), 'Matters relating to commitments – criteria for Joint Implementation, Comments from member states on criteria for Joint Implementation', A/AC.237/Misc.33 u. Add., Geneva.

Intergovernmental Negotiating Committee for a Framework Convention on Climate Change (1994), 'Matters relating to commitments – criteria for Joint Implementation, Comments from Parties or other member states', A/AC.237/Misc. 37, Geneva.

Intergovernmental Panel on Climate Change (1996), Climate Change 1995, *The IPCC Synthesis*, Geneva.

Intergovernmental Panel on Climate Change (1997), *The Regional Impacts of Climate Change*, Geneva.

International Council on Local Environmental Initiatives (1993a), *Klima schützen heißt Städte schützen*, Freiburg.

International Council on Local Environmental Initiatives (1993b), *Profiting from Energy Efficiency*, Toronto.

International Energy Agency (1998), *World Energy Outlook*, Paris.

Jepma, Catrinus (ed.) (1995), The Feasibility of Joint Implementation, Dordrecht.

Jochem, Eberhard, Heinrich Herz and Wilhelm Mannsbart (1994), *Analyse und Diskussion der jüngsten Energiebedarfsprognosen für die großen Industrienationen im Hinblick auf die Vermeidung von Treibhausgasen*, Bonn.

Klima-Bündnis (1994), *Lokale Politik für eine globale Herausforderung*, Frankfurt.

LeBlanc, Alice (1997), *An Emerging Host Country Joint Implementation Regime: the Case of Costa Rica*, New York.

Lelieveld, Jos and Paul Crutzen (1994), 'Emissionen klimawirksamer Spurengase durch die Nutzung von Öl und Erdgas', *Energiewirtschaftliche Tagesfragen*, (7) 435–40.

Ling, Chee Yoke (1994), 'A floundering GEF erodes confidence in green funding', *Third World Resurgence*, (42/43), 4–5.

Loske, Reinhard and Sebastian Oberthür (1994), 'Joint Implementation under the Climate Change Convention: Opportunities and pitfalls', *International Environmental Affairs*, (1), 45–58.

Lunde, Leiv (1997), 'Joint Implementation: a case study of Norway', ECON Report 16/97, Oslo.

McDonald, Susan (1994), 'Brazil's José Goldemberg: Exporting progress, importing change', *E-notes*, (2), 1–7.

McKinsey & Company, Inc. (1989), 'Background paper on funding mechanisms', in *Nordwijk Conference Report*, vol. 1, Annex, Noordwijk.

Matsuo, Naoki (1998), *Points and Proposals for the Emissions Trading Regime of Climate Change*, Shonan Village.

Matthes, Felix (1994), 'Necessary incentives for Joint Implementation projects from an investor's perspective', in Climate Network Europe (ed.), *Joint Implementation – from a European NGO Perspective*, Brussels, pp.50–59.

Michaelis, Hans (1992), 'CO_2-Minderung nach Rio', *Energiewirtschaftliche Tagesfragen*, 502–10.

Mintzer, Irving (1994), 'Institutional options and operational challenges in the management of a Joint Implementation regime', in Kilaparti Ramakrishna (ed.), *Criteria for Joint Implementation under the Framework Convention on Climate Change*, Woods Hole Research Center, pp.41–50.

Moomaw, William (1994), 'Achieving joint benefits from Joint Implementation', in Kilaparti Ramakrishna (ed.), *Criteria for Joint Implementation under the Framework Convention on Climate Change*, Woods Hole Research Center, pp.11–14.

New York State Energy Office (1994), *Draft New York State Energy Plan*, New York.

Niederl, Karl and Hans Schnitzer (1994), *Ökoprofit Graz. Nachhaltiges, innovatives Wirtschaften*, Graz.

Organization for Economic Cooperation and Development (1994), 'Urban energy management', draft, Heidelberg.

Orlebar, Edward (1994), 'Call for "carbon bond"', *Financial Times*, 1 June.

Parikh, Jyoti (1994), 'JI survey II: Supporting north-south cooperation', *UN Climate Change Bulletin*, (4), 5–6.

Porter, Gareth, Raymond Clémençon, Waafas Ofosu-Amaah and Michael Philips (1997), *Study of GEF's Overall Performance*, Washington.

Project Carfix (1995), *Joint Implementation: From Concept to Reality*, San José.

PUCO News (1997), 'PUCO approves CG&E financing of global environmental project',17 July, Columbus.

Rose, Adam (1992), 'Equity considerations of tradable carbon entitlements', in United Nations Conference on Trade and Development (ed.), *Combating Global Warming*, New York, pp.55–84.

Rose, Adam, Bradt Stevens, Jae Edmonds and Marshall Wise (1998), 'International equity and differentiation in global warming policy', *Environmental and Resource Economics*, (12), 25–51.

Sanhueza, Eduardo, Saar Van Hauwermeiren and Bert De Wel (1994), 'Joint Implementation: conditions for a fair mechanism', Instituto de ecologia politica, Santiago de Chile.

Sharma, Ravi (1993), 'Equity fundamental to collaboration', *Joint Implementation – an Eco special report*, 1–2.

Singer, Stephan (1994), 'Die Klimarahmenkonvention von Rio de Janeiro 1992 – der Weg und ihr Inhalt', Umweltstiftung WWF, Frankfurt.

Stahl, Karin (1994), 'Nichtregierungsorganisationen und internationale Organisationen: Partizipationsmöglichkeiten und Demokratisierungspotentiale am Beispiel der UN-Konferenz "Umwelt und Entwicklung"', in Rolf Hanisch and Rodger Wegner (eds), *Nichtregierungsorganisationen und Entwicklung*, Schriften des Deutschen Übersee-Instituts Hamburg no. 28, Hamburg, pp.237–53.

Swiss AIJ Pilot Program (1997), *Program Overview*, Berne.

Tietenberg, Tom, Michael Grubb, Axel Michaelowa, Byron Swift and Zhongxiang Zhang (1999), 'International rules for greenhouse gas emissions trading', UNCTAD/GDS/GFSB/Misc.6, New York.

TransAlta Corporation (1994), *Pursuing Greenhouse Gas Offsets*, Calgary.

Trexler and Associates (1994), *Carbon Offset Project Profiles*, Portland.

Trexler and Associates (1996), *Carbon Offsets as Environmental Mitigation: A First Regulatory Case Study*, Portland.

United Nations (1993), *Report of the Intergovernmental Negotiating Committee for a Framework Convention on Climate Change on the work of its 8th session held at Geneva from 16 to 27 August 1993*, A/AC.237/41, Geneva.

United Nations (1994), *Report of the Intergovernmental Negotiating Committee for a Framework Convention on Climate Change on the work of its 9th session held at Geneva from 7 to 18 February 1994*, A/AC.237/55, Geneva.

United Nations Conference on Trade and Development (ed.) (1992), *Combating Global Warming, Study on a Global System of Tradable Carbon Emission Entitlements*, New York.

United Nations Conference on Trade and Development (ed.) (1994), *Combating Global Warming – Possible Rules, Regulations and Administrative Arrangements for a Global Market in CO_2 Emission Entitlements*, New York.

United Nations Conference on Trade and Development (ed.) (1996), *Legal Issues presented by a Pilot International Greenhouse Gas Trading System*, New York.

UN Framework Convention on Climate Change (1997a), 'Activities Implemented Jointly under the pilot phase, First synthesis report on Activities Implemented Jointly', FCCC/SBSTA/1997/12, Add.1 and Corr. 1–2, Bonn.

UN Framework Convention on Climate Change (1997b), 'Activities Implemented Jointly under the pilot phase', FCCC/SBSTA/1997/INF.3, Bonn

UN Framework Convention on Climate Change (1997c), 'Kyoto Protocol to the United Nations Framework Convention on Climate Change', FCCC/CP/L.7/Add.1, Kyoto.

UN Framework Convention on Climate Change (1998), 'Activities Implemented Jointly under the pilot phase, Second synthesis report on Activities Implemented Jointly', FCCC/CP/1998/2, Bonn.

UN Framework Convention on Climate Change (1999), 'Activities Implemented Jointly under the pilot phase, Update on Activities Implemented Jointly', FCCC/SB/1999/INF.1, Bonn.

United States of America (1994), 'Statement of intent for bilateral development, cooperation and joint implementation of measures to reduce emissions of greenhouse gases between the government of the United States of America and the government of the Republic of Costa Rica', Washington.

United States Congress, Committee on Environment and Public Works (1994), 'The national action plan for global climate change', Joint hearing before the Committee on Environment and Public Works and the Subcommittee on Clean Air and Nuclear Regulation, United States Senate, 103rd Congress, 25 October 1993.

United States Department of Energy (1997), 'DOE project solicitation for USIJI projects', Internet URL http://www.ji.org/jinews/060697.shtml accessed 7 July 1997.

Vellinga, Pier and Roebijn Heintz (1994), 'Joint implementation: a phased approach' in Kilaparti Ramakrishna (ed.), *Criteria for Joint Implementation under the Framework Convention on Climate Change*, Woods Hole Research Center, pp.5–10.

Vos, Hans (1997), 'Business as usual?', Report for ECON, Amsterdam.

Watson, Robert (1998), 'Progress and challenges in mainstreaming the environment', *Environment Matters*, Fall, 6–9.

Wegner, Rodger (1994), 'Anspruch und Wirklichkeit in der nicht-staatlichen Entwicklungszusammenarbeit auf den Philippinen', in Rolf Hanisch and Rodger Wegner (eds.), *Nichtregierungsorganisationen und Entwicklung*, Schriften des Deutschen Übersee-Instituts Hamburg no. 28, Hamburg, pp.119–40.

Wexler, Pamela, Irving Mintzer, Alan Miller and Dennis Eoff (1995), 'Joint Implementation: institutional options and implications', in Catrinus Jepma (ed.), *The Feasibility of Joint Implementation*, Dordrecht, pp.111–32.
Yamin, Farhana (1993), The Climate Change Convention and Joint Implementation: Legal, Institutional and Procedural Issues, London.

2. Flexible Instruments and Stakeholder Interests: a Public Choice Analysis

Sandra Greiner[1]

INTRODUCTION

Given the decision at Kyoto on no fewer than four flexible instruments, and their dominating role in the debates, it might be thought that the international negotiating community puts a high priority on the efficiency of abatement measures, which, from a normative standpoint must, of course, be welcome. However, looking at the nature of efficiency as a *collective rationality* it has to be doubted that this really is the driving force behind individual action. Overall efficiency itself might well be desirable from an economic point of view, but surely it is of no particular concern to any individual party. Especially if interests are as diverging as they are in international climate politics it is not very probable that individual stakeholders in the negotiating process will seek to further the common good by introducing flexible instruments. Rather, they will pursue their own very specific interests, and the highly plausible efficiency argument seems to be only a vehicle for the promotion of these interests. The focus will therefore not be on the efficient use of flexible instruments but on how they can be designed in order to best serve individual interests.

In this chapter, the existence of four different mechanisms which all aim at promoting efficiency will be interpreted as the result of political bargaining in which different interest groups try to leave their mark on the mode of operation. Thus, by looking at the stakes involved and the most favourable design of flexible instruments for each group of stakeholders, we hope to explain some of the outcomes concerning flexible instruments. Special attention will be paid to the position of developing countries. Since the decision on flexible instruments does not involve social planning in any form, but is left fully up to the negotiating stakeholders, their ideal of how these

1. Hamburg University, Institut für Finanzwissenschaft, Von-Melle-Park 5, 20146 Hamburg, Germany, Tel.: +49-40/4123-5563, Fax.: +49-40/4123-6713, E-mail: greiner@hermes1.econ.uni-hamburg.de

instruments should be specified seems particularly relevant. To model stakeholder interests in the process, the study will adopt a public choice perspective which allows for some basic assumptions on interest positions.

THE MODEL

The theory of public choice focuses on the role of individual interests in the political process. Based on the assumption that individuals act selfishly not only in the economic sphere but also in the political one, the theory applies the famous *homo economicus* of economic theory to political decision making: as is assumed for economic agents, political agents are modelled as rational individuals who act to maximize their own utility (Mueller, 1989, pp.179ff). In this way, self-interest is the major force which determines political decision making. Thus the public choice approach takes an individualistic perspective by explaining political outcomes by means of individual behaviour.

For analytical purposes, three major groups can be distinguished which, in a representative system, shape the political process. Each individual belonging to one of these groups is assumed to share the group-specific interest position. (A good overview of public choice theory is given by Mueller, 1989; Pommerehne and Frey, 1979; Kirsch, 1993; for origins of the model, see Downs, 1957; Olson, 1965.)

Politicians are mainly interested in building up power, influence and prestige. Since their power depends on election, they choose political programmes so as to maximize their votes. The fundamental hypothesis is the Downsian argument that parties (as associations of politicians) 'formulate policies in order to win elections, rather than win elections in order to formulate policies' (Downs, 1957, p.28). Often, politicians are also acknowledged to have their own political preferences and values.

Interest groups are associations of individuals with relatively homogeneous tastes and incomes or who face the same situation. By investing in coordinated lobbying activities they aim at influencing politicians and administrative officials in order to promote their common goal.

Voters favour the political programme that best suits their individual interests. These interests are highly dependent on the economic gains or losses which are expected from the programme and vary according to the individual's situation, but voters may also have ideological preferences. However, being only able to vote for representatives, who are vaguely identified with their preferred programmes, they have very little influence over specific issues. Also, they are not fully informed about programmes and candidates, since gathering information may be more costly than the expected utility of voting for the right candidate, given the negligible influence of a single vote. This rational ignorance strengthens the position of interest groups, as politicians will be more aware of well-informed and homogeneous groups of voters.

The most important findings of public choice theory that will be used in this analysis are as follows.

1. Only individuals have interests, groups do not and neither do nations. What is often referred to as the national interest or a 'higher goal' is simply a political phrase, employed by politicians or interest groups to gain support for their own interests.
2. Political action must not be measured against what has been its stated intent but against the economic incentives being set.
3. 'Logrolling' is an individually rational method to reach one's goals: if there are different fields where concessions can be made in the political process, there will be bargaining and the exchange of concessions.
4. Politicians seek to take action where benefits are clearly noticed but costs are hidden. For this reason political decisions tend to benefit interest groups (which are relatively homogeneous and well-informed) and discriminate against the general public by spreading costs widely and making them unnoticeable.

However simple these assumptions may appear, they can well help us to understand some of the underlying mechanisms in political bargaining. By assuming that all political decision making is in fact interest-based, public choice theory very strongly questions the existence of general interests and concerns. In the international context where interests, values and situations differ even more than they do nationally, the idea of public interests is even more critical. The meaning of individual interests in the negotiations becomes particularly obvious. Thus, following the public choice approach, some insights may be gained into issues of supposedly general concern. The following analysis therefore builds on the assumption that neither protection of the earth's climate nor the efficiency of abatement strategies could by themselves be relevant issues unless individual stakeholders *find them* relevant and somehow connect them with their own utility.

APPLYING THE MODEL TO CLIMATE POLITICS: WHAT IS AT STAKE?

The discussion on cost-efficient abatement strategies will therefore be seen in terms of strategic interests. As responsible actors in the process, four groups of stakeholders can be identified which directly or indirectly shape the outcome of international climate negotiations and decide on the modalities of flexible instruments. Officially in charge of negotiating emission targets and the means of compliance are the governmental delegates of the participating states. In terms of interest positions, two opposing groups can be distinguished: on the one hand there are the governmental delegates of countries with targets, which are interested in undertaking emission reductions abroad (OECD countries that are listed as Annex B countries in the Kyoto Protocol); potential host countries, on the other hand, are those which either have not accepted any emission targets at all (the developing countries) or whose targets are not binding because business-as-usual emission scenarios lie below the agreed emission limitation, as is the case for most countries in transition. The governmental delegates of these countries will, of course, have other stakes in the use of flexible instruments than those of potentially investing countries.

Indirectly influencing the outcome of climate negotiations are the various non-governmental organizations (NGOs) which attend and observe the international conferences on climate change. Although not entitled to vote, these groups can by no means be considered to be without influence: by providing information both in and outside the conferences they are quite powerful behind the scenes, for example through person-to-person contacts with members of national delegations or in disseminating detailed studies on special issues. At the international climate conferences the actively involved NGOs can roughly be subdivided into two groups: business associations, on the one hand, and environmental and developmental NGOs, on the other (UNFCCC, 1996).

According to the assumptions of the public choice model, the actors involved in the negotiation process will pursue very specific interests according to the group to which they belong. As interest positions towards flexible instruments follow the more fundamental interests in climate politics these will be discussed first in order to deduce potential stakes in flexible instruments.

Governmental Delegates of Investing Countries (OECD Countries)

Like all politicians, the governmental delegates of investing countries will try to take action popular with their voters to ensure re-election. In this respect, climate policy as one issue among many others is only relevant if it captures voters' attention (Michaelowa, 1998, pp.252f). If we assume a correlation between economic wealth and the desire for environmental protection, voters' preferences for climate politics are relatively high in the investing countries, especially after meteorological extremes. Because of the high complexity of the subject, voters' information costs are also high. Visible and easily understandable measures to mitigate greenhouse gases should therefore be presented to the public. On the part of the governmental delegates there is a need for symbolic action (Gawel, 1995, p.61). The incurred costs, however, should be felt as little as possible by potential voters. Preferably, action against climate change should be taken by other countries or should be shifted into the future. At the conference level, the governmental delegates of investing countries will therefore try to ensure that all countries adopt emission targets and engage in climate protection activities. The accepted emission targets tend to lie in the distant future, long after the personal term, as for example the German 25 per cent target for 2005 accepted by the German Chancellor Kohl (Michaelowa, 1998, p.252). If costs cannot be shifted completely to other countries or into the future, they should at least be spread widely among voters, making them unnoticeable to any particular group. On the other hand, action against climate change may also open up opportunities to favour particular industries as the producers of environmentally sound technology, for example by supporting them in opening up new markets abroad.

In this respect, flexible instruments will offer some possibilities for the governmental delegates of investing countries. First, big emission trading deals bring spectacular publicity and are an easily calculable contribution to climate protection. Second, carrying out emission reductions abroad can reduce the need for national reduction programmes. These would incur more obvious – and presumably higher – costs to voters and possibly lead to losses in popularity. Third, special interest groups can be favoured through strategic trade policy: within the projects, national industries receive the opportunity to export technology and possibly establish long-term trade relations with the host countries. Since projects carried out abroad are highly opaque to the national public, this allows for extended support of the investing firms, for example by cosponsoring projects through developmental aid funds. (For a comprehensive discussion of the benefits of tied aid to the interest groups of donor countries, see Michaelowa, 1996.) An analysis of projects in the AIJ pilot phase shows that tying is already widespread (Michaelowa *et al.*, 1998).

The discussion of flexible instruments also enhances the investing countries' strategic position vis-à-vis the developing countries in climate negotiations.[2] In particular, bilateral relations between the host and the investing country may create dependence on the part of the host country since resources are being transferred within the project. If transfers are at stake and a bilateral relationship is established, host countries may vote with the industrialized countries in exchange for concessions and thus weaken the position of the developing countries. Also, the time spent on negotiating the modalities of flexible instruments may impede awkward discussions on climate responsibilities and the adoption of targets since conference time is limited.

2. Since the interest positions of developing countries and the industrialized countries are opposing in many respects, there has mostly been a clear-cut conflict line between the two sides. While the developing countries seek to hold the industrialized countries responsible for climate action and emphasize their developmental necessities, the industrialized countries like to include all countries in the process of climate protection.

There might however be some reservations regarding flexible instruments if governmental delegates fear moral criticism especially by environmental and developmental NGOs. The advantages of flexible instruments must then be weighed against the negative influence this criticism may have on voters, so that governmental delegates might not be able to reap fully the profits described above. It is unclear whether or not they will also have to fear the loss of national jobs due to the investment of national capital abroad. The negative effect on national jobs might be balanced or even outweighed by the effects of cost savings compared to a situation in which only domestic emission reduction measures are undertaken, and in which therefore even more jobs might be threatened due to higher costs.

Governmental Delegates of Host Countries

As one influencing factor, political power in the host countries depends on the amount of foreign transfers received, since these can be spent on diverse support-gaining activities. The governmental delegates of host countries will therefore try to combine the issue of climate protection with the receipt of additional transfers. The more freedom they are allowed in spending these transfers the better they will manage to meet their supporters' preferences and the more political advantage can be taken. The governmental delegates of host countries will therefore favour untied aid over project-oriented transfers. In contrast to the delegates of investing countries, they are not so much in need of presentable action against climate change, either because climate change is not a dominating issue among voters and could even be considered a 'spleen of the industrialised countries' (Michaelowa and Dutschke, 1998, p.18) – if we stick to the assumption of a positive correlation between income and environmental preferences – or because they can successfully claim that the industrialized countries should be held responsible for global warming. Emission targets should therefore be adopted neither today nor in the foreseeable future. If voters are more concerned with the issue of global warming because they live in countries which are particularly vulnerable to the effects of climate change (for example, the small island states), delegates will be interested in obtaining compensation payments from the industrialized countries to provide for adaptation strategies.

Considering these stakes, the delegates of host countries will have an ambiguous position towards flexible instruments. The following positive relationship can be identified.

1. Flexible instruments might be a vehicle for obtaining additional transfers if credits for Certified Emission Reductions partly accrue to the host countries and these can be traded. Credits can also be attractive if the host countries have to adopt binding targets in the long run and credits can be banked until then.

2. Framework contracts on bilateral joint implementation or CDM relations may lead to closer economic cooperation with the investing country and thus may help to improve the economic situation.
3. Within the projects, technological and financial resources are transferred from the investing country to the host country without costs for the latter.

On the other hand, flexible instruments might also have undesired effects. First, there is a possible competition between emission reduction projects and other, more favourable, forms of transfers. At worst, an increasing amount of transfers through flexible instruments might lead to a noticeable reduction of untied foreign aid. Most obviously, if flexible instruments became more important over time the industrialized countries might no longer be willing to fund sufficiently the GEF, which is a preferable mechanism because of its double majority voting system. Moreover, scarce human capital is used in the projects that could have higher returns elsewhere.

Second, there is a constant fear that the use of flexible mechanisms might in the long run serve as a vehicle for forcing host countries into the adoption of emission targets. Since host countries without targets face the moral hazard of increasing their emissions beyond the business-as-usual emissions if extra emission reduction possibilities can be sold, this so-called 'hot' or 'tropical air' problem (see Michaelowa, Chapter 1 in this volume) might serve as a good argument for the investing countries to carry out only projects with countries which have adopted binding targets. The delegates of least developed countries especially must fear that they will not receive their share of foreign transfers because they are least able to adopt targets.

Third, there is a fear that cheap reduction possibilities are being used up. If eventually host countries have to adopt targets, the marginal costs of emission reduction will be higher. To avert this danger, credits of emission reduction projects should at least be shared and host countries should be able to bank them.

Environmental and Developmental NGOs

NGOs which are motivated by either environmental or developmental issues, or even both of these, can always be considered to be defenders of a common good. Their identity builds on the struggle for non-individualistic goals such as the protection of the earth's atmosphere or the abstract notion of social equity. Identified as 'moral agents', their capability of attracting new members or funds rests on their commitment and credibility. They will be measured against the perceived influence exerted on political decision making. Furthermore, environmental and developmental NGOs will find it easier to gain support if the priority put on their respective goals is high. Thus they will not only strive to build up expert reputation in their particular field but they will also seek to promote public awareness of the issue on which they are campaigning.

In climate politics environmental NGOs will take a real interest in mitigating climate change. Their lobbying aim will be to raise public awareness of the climate issue both in and outside the conferences. At the conference level they will campaign for the adoption of binding emission targets which guarantee a sustainable level of greenhouse gases. Depending on ideological positions, views on how these targets could be fulfilled lie between two poles: 'deep' ecologists argue that this can only be done by a radical change in life styles and welfare criteria in the industrialized countries, serving as a blueprint for the developing world (see, for example, Climate Network Europe, 1994). More pragmatically oriented groups, however, trust in the total effect of a variety of individual reduction measures and in the effects of technological progress. As for NGOs with developmental goals, these will have very similar interests in climate politics to those of the politicians of host countries: developing countries should be excluded from the adoption of emission targets and should be compensated for environmental policies. In contrast to the host countries' politicians, however, developmental NGOs will put the emphasis on local benefits since these groups often have their roots in specific regions or fight for the preservation of some ethnic groups' living conditions (Michaelowa and Dutschke 1998, p.19). They will therefore favour small-scale projects benefiting the regional population rather than large amounts of untied foreign aid. Preferably, these projects should be carried out in cooperation with local developmental NGOs.

Considering these stakes, the employment of flexible instruments can have positive effects.

1. If the costs of abatement strategies can be lowered through the use of flexible instruments, potential investing countries will be more willing to adopt emission targets. Since environmental NGOs will be measured against the real outcome of climate negotiations, these targets are of great value in proving their successful operation.
2. Since efficiency works in both directions, the allowance of flexible instruments can be used to argue for emission reductions higher than those initially agreed upon.
3. Cooperation between the industrialized and the developing countries will promote the diffusion of 'green' technology.
4. As far as developmental interests are concerned, flexible instruments offer a transfer of financial and technological resources.
5. Environmental and developmental NGOs might actually be involved in the projects carried out under the flexible mechanisms. NGOs might serve in the verification, monitoring or certification process of the projects. This not only gives them the chance of building up a reputation as experts in

practical environmental protection activities but it potentially also offers additional sources of income.

On the other hand, environmental and developmental NGOs must also fear adverse effects on their particular interest position. First, there is a constant danger that the actual emission level will be higher if flexible instruments are employed, either because project results tend to be overestimated if monitored poorly, since all project participants will be interested in the creation of a large amount of credit, or because the existence of 'hot' or 'tropical air' in the baseline scenarios of non-OECD countries can be used in a manipulative way. Second, used as a strategy, the discussion on flexible instruments draws attention away from more important issues, such as the adoption of emission targets or the domestic implementation of climate policy. Third, some projects, such as the creation of monocultures are subject to criticism for ecological reasons. Finally, if the selection and conception of the projects is mainly in the hands of the investing partner, there is the danger of more 'white elephant' projects which do not entail benefits for the local population. Then the potential resource competition between flexible instruments and the GEF becomes a major threat.

Business Associations

Under the assumption of profit maximizing behaviour, business associations representing the interests of specific business branches will strongly oppose binding emissions limitations for their members because these would only incur additional costs or restrictions. With the exception of some 'green' sectors,[3] they will therefore lobby against the adoption of binding targets. For example, the potent US emitters' lobby 'Global Climate Coalition' spent up to 13 million dollars on advertisements against the Kyoto Protocol, stressing its negative impact on jobs (Toman *et al.*, 1997, p.10). If emission cuts cannot be avoided, though, they should at best be state-subsidized, followed by voluntary agreements. Also these associations will lobby for 'grandfathered' permits rather than taxes or auctioned permits.

Although the issue of climate politics will not be popular with most business associations, flexible instruments will be regarded positively. If emissions limitations cannot be avoided in the first place, flexible instruments will help to reduce compliance costs. Also the opportunity of gaining access to new markets abroad as a by-product of the projects adds further appeal to the use of flexible instruments. At best, already existing business relations involving energy-saving components could be turned into

3. Despite their growing significance at international conferences, these lobbies, mainly representing the interests of the producers of renewable energy technology and of reinsurers, will not be treated here because they do not have a specific position on the issue of flexible instruments.

certifiable joint implementation or CDM projects. In the past, business associations or individual companies have not only declared their positive interest in flexible instruments but have also actively supported their evolution, as for example in the case of Chevron, which sponsored a 1998 CDM workshop in Africa (Michaelowa, 1998, p.255).

Stakeholder Views on How Flexible Instruments should be Put to Work

Going back to our initial statement that individual actors involved in the negotiation process of flexible instruments will not care about the efficient design of these instruments but will try to adjust the modalities of their operation to their individual interests, we can now be more specific about their preferred designs. How individual stakeholders will try to shape the design of flexible instruments in the negotiations will predominantly depend on their individual stakes. Not surprisingly, stakeholders with rival interests in climate politics will also favour opposing implementation modes of flexible instruments.

Looking at the stakes involved, we will suppose that business associations will lobby for a maximal range of application of flexible instruments and oppose all kinds of regulation. Emitters should be granted maximum freedom in choosing project types and partners so that the projects can be best adjusted to company-specific profit opportunities. Environmental and developmental NGOs, on the other hand, will emphasize possible misuses caused by rent-seeking behaviour. They will therefore demand tight regulation standards and efficient monitoring precautions, preferably under NGO supervision. Added to this, emission reductions achieved through flexible instruments should automatically be discounted to take account of the expected overstatements and to appropriate part of the efficiency gain for ecological purposes. Moreover, deep ecologists will argue for a certain ratio of emission reductions to be reached domestically in the investing countries.

The politicians of investing countries and the politicians of host countries will disagree on the question of credit sharing between host and investor, among others. While the politicians of investing countries will argue that all the credits should accrue to the investor, the politicians of host countries will demand their share. Given that host countries are not in need of credits for meeting their national targets today, they will mainly be interested in trading the credits acquired for money. All countries should therefore have access to an international trading system independent of the adoption of national targets. If national targets are expected to be binding in the future, credits should be bankable without discounting. Even more favourable than credits would be direct side-payments as a fixed ratio of cost saving achieved by the projects.

The politicians of the investing countries will, however, oppose all forms of open or hidden transfers. Preferably, the issue of flexible instruments should not be mixed at all with side-payments or transfers other than that which will be transferred within the projects. On the other hand, they will take advantage of strategic potentials if possible. For example, they will be interested in excluding countries without targets from an international trading system in order to force these countries to adopt targets. It is unclear whether or not they will favour the discounting of credits acquired in countries without targets. On the one hand, the discrimination of these countries might increase the pressure to adopt targets, but on the other, the investing country will have to pay a higher price for emission reductions since discounting credits raises the marginal costs of the projects (Michaelowa and Dutschke, 1998, p.11).

Against the interests of environmental NGOs, the politicians of both host and investing countries will try to appropriate the rent of 'hot' or 'tropical air'. For example, the politicians of host countries could turn the difference between actual emissions and the higher emission target into tradable permits. They will then have an interest in emission reductions acquired by national 'efforts' also being credited. The politicians of investing countries could get hold of the rent by formulating joint targets with non-OECD countries. Since appropriation will be easier if the non-OECD country cannot market its hot or tropical air on its own, again no country should in their view be allowed participation in the trading system if no national targets have been accepted.

Table 2.1 summarizes interest positions towards flexible instruments and the ideal modalities of their implementation for each group of stakeholders.

Table 2.1 Optimal design of flexible instruments for different stakeholders

Actors	Stakes in flexible instruments	Optimal design
Governmental delegates of investing countries	Presentable and easily calculable action against climate change Reduction of national costs Strategic trade policy and favouring of special interest groups Source of strategic power in climate negotiations Fear of moral criticism	All credits should accrue to the investing entity The employment of flexible instruments should not be tied to other issues, no conditions concerning side-payments should be imposed Projects should be conceptualized on a bilateral basis between host and investing country to allow for strategic action Countries without binding

		targets should not be allowed trading of reduction credits Restrictions on the use of flexible instruments should only be imposed if there is open criticism by NGOs
Govern-mental delegates of host countries	Receipt of credits at no expense to the host country. These are of interest if credits can be banked and binding emission targets are expected to be within reach, or credits can be traded Closer economic cooperation with the investing country Transfer of financial and technological resources But: Possible decrease of more favourable forms of transfer Feasr of being forced into the adoption of binding emission targets Exhaustion of cheap reduction possibilities	Credits for emission reductions should be equally distributed between host and investor; and/or other forms of side-payments should be linked to the mechanisms Credits should be bankable All countries should be allowed to buy and sell tradable permits, independent of the adoption of emission targets Emission reductions achieved domestically should also qualify for crediting Projects should be evaluated and approved on a multilateral basis such as with a clearinghouse to avoid strategic behaviour on the part of investors
Environ-mental and develop-mental NGOs	Countries will be convinced more easily to adopt targets if abatement costs are lowered Flexible instruments could be used for higher reduction claims Diffusion of 'green' technology Transfer of financial and technological resources Direct participation possibilities But: Moral hazard of exaggerating project results Some projects may be	Projects must be approved and monitored by NGOs Risk discounting of projects Prescribed ratio of domestic emission reductions External benefits of projects must be guaranteed

	environmentally damaging Danger of 'white elephant' projects	
Business associa-tions	Cost reduction if emissions have to be lowered Chance of establishing new business relations	Only investors conceptualize projects Credits accrue fully to the investor No regulations

CONCLUSIONS

From a public choice perspective, the decision on four different mechanisms laid down in the Kyoto Protocol, all of which centre around the idea of cost-efficient abatement strategies, can be explained as the result of diverging interests. The advantage of four such mechanisms over only one lies in the fact that more interests were able to be considered. Where interests are strictly opposing and hard to reconcile the creation of different systems seems the only way out. Especially the diverging interests of host countries and investing countries seem to have split the scenery in half. While the CDM as the relevant mechanism for the developing countries is very close to what the politicians of host countries would consider optimal, placing emphasis on the developmental rather than on the ecological issue, the provisions concerning joint implementation between Annex I countries seem to be more in line with investors' interests as described in Table 2.1. Contrary to the numerous precautions of the CDM, joint implementation between Annex I countries only needs to be approved by the parties involved and to be additional to any reduction in emissions that would otherwise occur (Art. 6). Thus projects can be negotiated and carried out freely on a bilateral basis. The revenue accruing to the partners is maximal, since no percentage is diverted to third parties. The CDM, on the other hand, tends to meet the interests of host countries' politicians and NGOs. Article 12 explicitly reserves part of the projects' revenues for countries especially vulnerable to the adverse effects of climate change and clearly envisages a multilateral fund (Michaelowa and Dutschke 1998, p.19). Also the possibility of credit banking until the first commitment period is outlined.

The creation of different flexible mechanisms which favour either group of stakeholders might, however, only be a short-term solution. If the demand for and supply of the projects diverge, some stakeholders will become more powerful, giving them the chance of imposing their ideals of flexible instruments on others. Possibly there will be a shortage of capital dedicated for projects through the CDM. Since investors will favour the more liberal conditions of joint implementation with countries in transition they might shy away from the CDM as long as the marginal costs of abatement measures are not considerably lower in the developing countries and as long as there is no shortage of project opportunities in the transition countries. Developing countries might then be deprived of the benefits of additional transfers. Also, if the developing countries are excluded in the long run from the international trading system, their solidarity may shrink. Just as at the beginning of the debate, when the developing countries altogether rejected the concept of joint implementation, they might be forced into reconsidering their positions because individual countries might change sides and adopt national targets, attracting more than their share of foreign investments.

REFERENCES

Climate Network Europe (ed.) (1994), *Joint Implementation from a European NGO Perspective*, Brussels.

Downs, Anthony (1957), *An Economic Theory of Democracy*, New York.

Gawel, Erik (1995), 'Zur Neuen Politischen Ökonomie der Umweltabgabe', in Wolfgang Benkert, Jürgen Bunde and Bernd Hansjürgens (eds), *Wo bleiben die Umweltabgaben? – Erfahrungen, Hindernisse and neue Ansätze*, Marburg, pp.47–101.

Kirsch, Guy (1993), *Neue Politische Ökonomie*, 3. edn. Düsseldorf.

Michaelowa, Axel (1998), 'Climate policy and interests groups – a public choice analysis', *Intereconomics*, 33(6), 251–9.

Michaelowa, Axel and Michael Dutschke (1998), 'Creation and sharing of credits through the Clean Development Mechanism under the Kyoto Protocol', HWWA Discussion Paper no. 62.

Michaelowa, Axel, Katharina Michaelowa and Scott Vaughan (1998), 'Joint Implementation and trade policy', *Außenwirtschaft*, 53(4), 573–89.

Michaelowa, Katharina (1996), 'Who determines the amount of tied aid? A public choice approach', HWWA Discussion Paper no. 40.

Mueller, Dennis C. (1989), *Public Choice II*, rev. edn of *Public Choice*, Cambridge.

Olson, Mancur (1965), *The Logic of Collective Action*, Cambridge, Mass.

Pommerehne, W.W. and Bruno S. Frey (eds) (1979), *Ökonomische Theorie der Politik*, Berlin–Heidelberg–New York.

Toman, M., M. Tebo and M. Pitcher (1997), *A Summary of US Positions on Climate Change Policy*, Washington, Resources for the future.

UNFCCC (1996), *WHO IS WHO in the UNFCCC Process 1995–1996*, edited by the Secretariat of the United Nations Framework for Climate Change, Bonn.

3. Making the Clean Development Mechanism Compatible with the Kyoto Protocol

Catrinus Jepma and Wytze van der Gaast

INTRODUCTION

In December 1997, the third conference of the parties (CoP3) agreed upon a protocol to take appropriate action for the period beyond 2000 to achieve the objectives of the Framework Convention on Climate Change (FCCC). In the framework of this, Article 3 of the protocol includes a provision that Annex I parties 'shall, individually or jointly, ensure that their aggregate anthropogenic carbon dioxide equivalent emissions of the greenhouse gases listed in Annex A do not exceed their assigned amounts, calculated pursuant to their quantified emission limitation and reduction commitments inscribed in Annex B [to the protocol]' (Article 3.1). The total assigned amount of all Annex I parties is to be at least 5 per cent below the emission levels of 1990 in the commitment period 2008 to 2012. Since CoP3 was held in Kyoto (Japan) the protocol is referred to as the 'Kyoto Protocol'.

Among the several arrangements that the parties have agreed to under the protocol is the decision that Annex I parties may use the so-called 'flexibility mechanisms' in order to achieve their commitments under Annex B in a cost-effective way. These instruments are the following.

- Joint implementation (JI) among Annex I parties, which is defined in Article 6 of the protocol. This instrument allows an Annex I party to transfer to or acquire from other Annex I parties emission reduction units resulting from greenhouse gas emission reduction projects.
- The clean development mechanism (CDM) which is defined in Article 12. The CDM envisages the establishment of a multilateral mechanism the purpose of which is to assist non-Annex I parties in achieving sustainable development and to assist Annex I parties in achieving compliance with their commitments under the protocol. The

cooperation under the CDM is rather similar to the concept of JI between Annex I parties and non-Annex I parties where through project-based cooperation non-Annex I parties are assisted in their sustainable development and Annex I parties may obtain credits in the form of certified emission reductions.
- International emissions trading (IET). Article 17 of the protocol defines the opportunities for Annex I parties to participate in emissions trading for fulfilling their protocol commitments.

With respect to these mechanisms, several issues still have to be resolved before they can become operational. At the fourth meeting of CoP in Buenos Aires (November 1998) a lively debate on the implementation of flexibility mechanisms took place. Eventually, CoP4 agreed on a work programme that would have to be completed by 2000. This work programme contains issues related to the implementation of joint implementation, the establishment of an international trading system and the setting up of the clean development mechanism (CDM). It was noteworthy to notice that especially developing countries – who have always opposed joint implementation as an instrument to allow industrialized countries flexibility in achieving their objectives and commitments under the FCCC – showed a great interest in the CDM. It seems that they have embraced the CDM as an instrument now that the CoP has decided that the GHG emission reductions from cooperative projects between industrialized and developing countries can only be certified if the project assists the developing host countries in their efforts to follow a sustainable development path.

Article 12 of the Kyoto Protocol, however, does not describe the CDM structure very well; it only provides a framework that needs to be completed before the CDM can become operational. In this chapter some of the questions that have come up since Kyoto will be addressed. These questions have, among other issues, been included in the work programme on flexibility mechanisms adopted by CoP4. The questions related to the CDM that will be addressed in this chapter are the following.

1. How to determine the GHG emission reductions achieved through a CDM project? Not only should the emissions of the new plant under the project be measured, also a reference scenario ('baseline') needs to be determined of what the emissions would have amounted to in the absence of the project.
2. Are forestry projects allowed under the CDM? In its current form, Article 12 only refers to projects that result in emission reductions, but some

observers have argued that this does not necessarily exclude 'sinks' projects.

3. What will the governance structure of the CDM look like? Article 12 envisages the establishment of an executive board for the CDM that could be assisted by so-called 'operational entities', but it is not yet clear which institute will do what and who will be responsible for certification procedures.
4. What exactly will be the amount that parties involved in CDM projects have to pay to cover administrative expenses and adaptation investments in the developing countries most vulnerable to climate change (Article 12.8)?

The concluding section addresses the question of the extent to which the CDM will be compatible with joint implementation and international emission trading in providing a portfolio of cost-effective investment options for industrialized countries to achieve their protocol targets.

BASELINE DETERMINATION: A GENERAL OVERVIEW

Already since the inclusion of JI in the FCCC in 1992, the issue of additionality of projects has been an important one. In order to determine what the greenhouse gas emission reduction or sequestration of a JI project is, it is necessary to determine a reference scenario to estimate what the emissions at the project site would have been in the absence of the project. The main difficulty with determining such a reference scenario (often referred to as the *baseline*) is that it is counterfactual. For example it describes a situation that will, because of the project, never exist. As a result, many have argued that, because of this, the additionality issue is the weak point of JI. Since parties may have an incentive to inflate the baseline so that a higher emission reduction can be claimed, a careful (third party) check is required to judge whether a project's baseline is correct and fair. The discussion on this is not finished yet, and will continue at the future sessions of CoP and CoP/MoP, (meeting of the parties).

Several options for baseline determination have been proposed in the literature. The fundamental point in this respect seems to be how one wants to consider the essential characteristics of a baseline. On the one hand, one may argue that the baseline needs to be a technically precise as possible description of the counterfactual situation of a particular JI project. This approach requires detailed information about the conditions under which a particular project is undertaken, and so on.

On the other hand, one can take the opposite view by arguing that, irrespective of the amount of detailed information gathered to construct a baseline, one will, to a certain extent, always have to make bold assumptions to construct the baseline. In other words, baseline determination will to a certain extent be an arbitrary and, therefore, subjective process. According to this approach, the particular characteristics of a project are therefore not extremely relevant. What matters, though, is what a *reasonable* baseline could be in a situation that is broadly comparable to the circumstances of the project at hand. In the following, and against the background of the remark made above, four different approaches to baseline determination will be discussed in somewhat greater detail, thereby shifting from primarily the first approach to predominantly the second.

Project-specific Baselines

The first option for baseline assessment is straightforward and deals with an ex ante best acceptable estimate of what the emissions on the site would have amounted to in the absence of the project. This estimate can be made in several ways and depends on the characteristics of the project and the host country where the project is to be implemented. We will not elaborate in great detail the many complexities (for instance, with respect to project boundaries, incorporation of externalities, and so on) that will arise in this approach, because this has been dealt with extensively elsewhere.[1] It may be illuminating, though, simply to illustrate how difficult it is to determine what the ex ante baseline is, even if the AIJ/JI/CDM project seems to be rather straightforward.

Just to give a small illustration, let us take, for example, a number of pilot energy efficiency improvement projects carried out in the Baltic region under the auspices of the Nordic Council of Ministers. According to the report compiled by the Council, most of these projects turned out to have only speeded up the investments.[2] Without the projects, the investment would probably have been made by the host countries anyway, but with a delay of three to five years. If these projects were JI projects under the Kyoto Protocol, the projects' baseline would only deviate from the actual emissions (for example, result in credits) during the first three to five years of the projects. It should be noted, however, that this conclusion probably does not hold for all countries in central and eastern Europe. Some of them have a better developed infrastructure and have achieved a higher income and

1. For example, see Chomitz (1998).
2. Nordic Council of Ministers (1997).

welfare level than other countries with economies in transition. The projects studied by the Nordic Council of Ministers are mainly implemented in countries that belong to the first category. For countries belonging to the latter category it will probably take (much) longer before they are able to carry out the investments themselves.

This example clearly illustrates the complexity of baseline determination. Some of the potential host countries are undergoing a process of rapid economic transition (as, for example, in most of the countries in central and eastern Europe or some rapidly growing developing countries). In these countries, several JI (or CDM) projects probably only speed up investments that would have been carried out anyway in the medium term. For several other potential host countries (for example, lower income developing countries), it is less likely that the JI (or CDM) project investment would have been carried out anyway in the short or medium term. For these countries, the period for which the JI (or CDM) project is additional is often (much) longer.

Determining the length of the period during which a JI (or CDM) project is additional is not the only uncertainty surrounding the baseline determination: factors such as economic growth, energy prices, currency prices and political risks can also be important. The difficulty is that if the project developers have determined a project baseline that indicates an additional emission reduction resulting from the JI (or CDM) project over a period of ten years, but after five years it turns out that the host country would obviously have carried out the project itself (for example, because the host country itself invests in several similar projects), the reported emission reduction is larger than what has actually been achieved. Such a case is obviously beneficial for the investing country and could be advantageous for host country parties, but is disadvantageous for the global climate.

Ex Post Baseline Corrections

A second option for dealing with the baseline issue is to follow the approach just mentioned but to allow for ex post corrections of the baseline. Such corrections may be required if it turns out that the underlying assumptions of the reference scenario were not correct. For investors, but also for the host country parties, this may increase the risk of investing in JI (or CDM) as it will not be clear beforehand how many credits the project will generate. On the other hand, ex post corrections of the baseline have the advantage that the generated credits are, most likely, based more on real emission reductions than would be the case without ex post corrections. In this case, as far as the project is concerned, it is not the global climate that runs the risk of losing, but the project partners themselves.

Were CoP to decide, if necessary, on ex post corrections of baselines, the project developers would probably tend to select only those projects which seemed extremely unlikely to be carried out by the host countries themselves, even in the medium term. In this respect it is worthwhile mentioning that, on the basis of a detailed analysis of 30 projects in central and eastern Europe,[3] it was concluded that three factors in particular hamper the automatic improvements in the processes of energy production and consumption in the region: (a) the funding available for emission reduction investments in power plants and district heating plants is often insufficient; (b) in several central and eastern European countries the legislation prescribing energy efficiency improvements is often lacking; (c) the technical and management skills to implement and maintain new energy-efficient technologies are often insufficient.

With respect to this, the above-mentioned analysis by the Nordic Council of Ministers[4] makes a distinction between projects on the energy demand side (for example, district heating) and those on the supply-side (for example, power plants). First, energy supply side investments are often much greater than demand side investments, which makes it relatively easier to invest in district heating improvements. Second, as a result of the gradual reduction of energy subsidies during the transition process, there is a larger pressure on governments to improve the energy efficiency on the energy demand side. Finally, consumers in central and eastern European countries are becoming more and more eager to have comfortable living conditions, including a an adequate domestic heating system. On the basis of this analysis, the Nordic Council of Ministers concludes that there is greater pressure on the governments to invest in projects on the demand side of the energy market than to invest in projects on the supply side. As a result, in the case of ex post corrections of baselines, JI energy supply-side projects in central and eastern Europe are probably less risky than demand side projects, since the baseline for supply side projects is probably more stable.

---

3. CCAP/SEVEn (1997); Van Harmelen *et al.* (1997).
4. Nordic Council of Ministers (1997).

Box 3.1 Costa Rican Protected Areas Project

An example of a project for which a methodology has been developed to deal with ex post baseline corrections is the Costa Rican Protected Areas Project.[5] This project aims at sequestering 15.6 million tons of carbon equivalent on an area of 530 000 ha. Through an international verification and certification procedure carried out by the Oxford-based Société Générale de Surveillance (SGS),[6] the government of Costa Rica has been able to issue a certificate (a so-called 'certified tradable offsets', CTO) for the first one million tons of carbon sequestered via the project. In order to ensure the buyers of CTOs a risk free (98 per cent covered) offset, 700 000 tons of carbon have been retained in buffer. According to the project developers, the largest component of the coverage of the buffer relates to uncertainty about the position of the baseline. They expect this uncertainty to correspond to 16.1 per cent of the total amount of carbon sequestered.

Top-down Baselines

A third approach for baseline determination, developed by the US Center for Clean Air Policy, was recently added to the debate: the methodology of top-down baselines. The idea is that national governments of JI host countries would use their quantified emission limitation or reduction commitment (QELRC), or, which boils down to the same, the assigned amount that follows from this restriction, as a basis to calculate for their respective sectors and/or technologies what the GHG emissions per unit of energy used would be at which their commitment could be fulfilled. So, to give an example, it might be that the QELRC of a central and eastern European party can only be achieved if, as a part of a whole set of measures, the CO_2 emissions per unit of energy produced in the power sector would be, say, 20 per cent less than the average in the present situation. In that particular case, the 20 per cent figure would then determine the baseline for JI projects in that particular sector, and so on.

With respect to CDM projects, a similar top-down methodology cannot be applied, simply because the non-Annex I parties will be the host countries

5. This example is included in this section for illustrative reasons. As will be discussed later in this chapter, it is still unclear whether forestry will be eligible for CDM and whether forest conservation projects, such as the one described here, will be eligible for JI.
6. For a description of this project, see Foundation JIN (1998b, pp.10–11).

here, and they have not accepted QELRCs. A similar norm for baseline determination to that suggested for projects in central and eastern European Annex I parties cannot, therefore, be applied for the non-Annex I parties. To solve this dilemma, it has been suggested that an attempt nevertheless be made to construct baselines on the basis of acceptable *simulated* targets for potential non-Annex I parties (which is obviously a politically tricky affair). The latter element of the top-down approach is indeed rather contentious, which made Goldberg (1998) once remark that this could create 'tropical air' in the determination of CDM projects' baselines.

A Baseline Default System

A fourth option, which was proposed for baseline determination quite recently, notably by Iestra, Jepma and Michaelowa,[7] is to adopt default project/technology-specific baselines with a possible differentiation per country or region. A panel of experts could determine a baseline for a number of project types, which could serve as a benchmark for the FCCC. This project categorization could then be extended to a categorization by regions or countries, resulting in a region-by-project matrix. Thus a *matrix of baselines* can be constructed, which project developers can consult. If an investing and a host country party agree on a project, they can just look up the baseline in the matrix and calculate the credits. An example of what the elements of the matrix may look like can be found in Michaelowa (1998).

There are a number of advantages to this option. First, the transaction costs for the project developers will be lower, as they probably do not have to hire consultants any more, or at least to a much lesser degree. A visit to, for example, the FCCC Internet home page may be sufficient. Second, a third party check for each individually determined baseline is no longer necessary, which may also result in a significant cost saving. A third advantage of such a categorization system is that it provides a way out of the dilemma of choosing the correct baseline out of several possibilities each of which can equally well be defended as correct.

One could argue that the matrix approach can be too imprecise, because in particular circumstances the matrix elements are so clearly unfair to the project participants that an ad hoc adjustment seems to be imperative. Therefore, as an additional element of this matrix approach, it has been suggested that the possibility be included for the project participants to appeal for an adjustment of the baseline used in their particular case. This

---

7. Foundation JIN (1998a).

opportunity would be optional: in other words, project participants can decide for themselves whether to take the risk of losing the appeal, by making an investment in data gathering in order to apply for an exemption. The extra costs associated with this procedure – the costs associated with the appeal – as well as with a possible extra third party verification, will have to be borne by themselves.

With respect to the procedure to set up the matrix system just mentioned, it has been suggested that FCCC authorized international third parties be allowed to participate in the process of determining the aggregate sector/technology set of baselines. Furthermore, a periodic international verification process for the aggregate baselines would be necessary insofar as technological progress would require this. A particular point in this respect is the risk of leakage between sectors: setting a target for one sector in a non-Annex I party may affect the appropriate target for other sectors, and so on.

ROLE OF SINKS

Comparing the texts of the Kyoto Protocol articles on project-based cooperation, a difference is found for the wording on the type of projects that can be carried out as JI and CDM. Whereas Article 6 explicitly allows the transfer or acquisition of emission reduction units resulting from both (JI) projects aimed at emission reduction or sink enhancement, Article 12 only refers to (CDM) projects resulting in certified emission reductions. With respect to this an important question arises: whether Article 12 implicitly includes sink enhancement projects (for example, through forestry) or excludes these. Whether or not sinks are, or will be, eligible for CDM project cooperation may be a crucial question for several non-Annex I parties (for example, Costa Rica and Brazil) seeking to become involved in CDM cooperation. This section offers some thoughts on the potential role of sinks in CDM projects given below.

Answering the question whether Article 12 implicitly includes or excludes sink enhancement projects leads us to the text of Article 3, which defines the QELRCs and the procedures for Annex I parties to meet these. The role of sinks is described in paragraph 3 of Article 3. According to this paragraph, parties may include in their emission inventories the net changes in GHG emissions by sources and removals by sinks. The latter includes forestry activities, which in Article 3.3 are limited to afforestation, reforestation and deforestation (of course in terms of emission increase). As this text directly links the QELRCs of Annex I parties to forestry activities, a direct link between emission reduction (or limitation) and sinks is established in Article 3.

On the basis of this reasoning, it could be argued that for articles *derived from* Article 3 the same direct link between emission reductions and sinks holds. After all, except for not including it in Article 12, there is no part of the Protocol which says that for international cooperation projects there should be no link between sink enhancement and emission reductions. Therefore, if CoP6 (when completing the work programme on flexibility mechanisms adopted at CoP4) decides that sink projects should not become eligible for CDM, it may be better to say so explicitly in Article 12 through an amendment. Otherwise, the formulation used in Article 3 on the role of sinks in parties' emission reduction may as well hold for Article 12 on CDM projects.

The ambiguity on this topic has led to some different country opinions and a request by the Subsidiary Body for Scientific and Technological Advice (SBSTA) and Subsidiary Body of Implementation (SBI) to the Intergovernmental Panel on Climate Change (IPCC) to investigate further the modalities of forestry projects in a special report, as several parties have expressed the wish to obtain more information on methodologies for baseline calculations, monitoring and verification, and ensuring sustainability of forestry projects.

The perceived lack of information on methodologies for forestry projects may explain why forestry is not explicitly mentioned in Article 12. Several parties have said that they consider the inclusion of sinks in the CDM at the present stage as risky. These parties fear that it may be relatively easy to inflate the Annex I parties' overall assigned amount if the methodologies for forestry projects are not clear. This may, of course, also hold for forestry projects in central and eastern Europe, but the risk of inflation is considered to be smaller, since the emission sequestration will be included in the eventual emissions inventory of the Annex I host countries. The latter is dealt with in Article 3.3 of the Protocol.

With respect to forestry and the CDM, some parties, among them the EU, have taken the position that in the short term forestry projects should not be eligible for the CDM until the methodological issues concerning sink-enhancing projects are solved: for example, on the basis of the above-mentioned IPCC special report. This position may, however, be a trifle arbitrary. From the discussion in the previous section it has become clear that for several other types of potential CDM projects a risk of baseline inflation also exists. These projects, in general, do not face the opposition that usually accompanies forestry projects. Moreover, for at least the last decade, considerable experience has been gained in implementing, monitoring and verifying forestry projects, which could be useful now for establishing forestry baselines. By way of precaution it may therefore be advisable not to let the public's generally negative perception of forestry lead policy makers to reject it as an option for CDM. Although careful consideration of its applicability is justified, one should not throw the baby away with the bath water.

Special List of Projects

Another option for dealing with the role of sinks enhancement, and other projects for which it may be difficult to determine the emission reduction, could be to limit the types of projects eligible for CDM to clean technology projects resulting in direct and relatively easy to verify emission reductions. In addition to such a list, a second list of projects could be considered. This list could contain transfers, whose contribution to greenhouse gas emission reduction is more difficult to determine. Such a 'special list of projects' could contain projects such as sink enhancement investments, but also projects supporting capacity building in developing countries.

Basically, the special list should have the same status as the main list of CDM projects: that is, special list projects should result in certified emission reduction, just as with regular CDM projects. The difference is that for projects on the special list no project-specific baseline will be determined. Credits could be awarded here as a lump-sum.[8] CoP could establish a system which, on the basis of 'best professional judgement' determines for each type of special list project the credits per unit of investment. A sink project in a particular region, for example, would result in a predetermined amount of credit per hectare, irrespective of the type of forestry, which could be a disincentive for those who aim at acquiring quick credits via 'plantation' forestry projects. An investment in environmentally sound capacity-building, to give another example, could result in credits per dollar invested, and so on.

A complication with determining lump-sum credits for special list projects is that the net emission reductions resulting, for example, from a capacity building project will possibly only take place after the first commitment period. A considerable part of these emission reductions may be achieved after that. This aspect would have to be incorporated in the credit determination. Another complication of the crediting of special list projects is

8. Besides a lump-sum crediting system, one could also envisage a crediting system which calculates, as far as possible, the emission reductions via a baseline determination, as discussed in the previous section, but subsequently discounts these credits owing to the higher degree of uncertainty. The discount factor could be set by CoP.

that the emission reduction may be real, but may not be directly measurable. According to Article 12.5(c), emission reductions through CDM projects can only be certified as 'credits' if they result in real, measurable, and long-term benefits related to the mitigation of climate change. A lump-sum crediting system for special list projects may well result in real benefits, and, if so, in long-term benefits as well. The difficulty lies in the measurability of the emission reduction. Also in this respect it seems to be an important consideration for CoP, whether a system developed on the basis of 'best professional judgement' is acceptable or that a detailed as possible estimate should be made of each emission reduction claimed.

Finally, it should be underlined that establishing a special list of CDM projects may be beneficial for the developing countries, as the projects on the list could contribute strongly to their sustainable development.

GOVERNANCE AND THE CDM

In terms of governance, Article 12 envisages an extensive institutional structure for the CDM. This structure contains, in hierarchical order, CoP as the Meeting of the Parties to the Kyoto Protocol (CoP/MoP), the CDM executive board and the 'operational entities'. Article 12 states that the CDM shall be subject to the authority and guidance of CoP/MoP, and that it shall be supervised by the executive board (Art. 12.4). The CoP/MoP will designate operational entities that will be responsible for the certification of the emission reductions, resulting from each project (Art. 12.5).

On the basis of Articles 6 and 12, it may be clear that the governance requirements for CDM are much stronger than for JI. For JI, the governance structure is mainly discussed in Articles 5 and 7, which deal with inventory requirements and reporting procedures for Annex I parties. For the non-Annex I parties participating in CDM projects such procedures are, however, not included in the Kyoto Protocol. In addition, since the focus of CDM projects is on sustainable development in the non-Annex I host country parties, a special certification procedure is required. All this has resulted in a text for JI, indicating that CoP/MoP1 *may* further elaborate on guidelines for the implementation of JI (for example, guidelines for baseline determination, monitoring and verification), whereas the text of Article 12 states that, with respect to the CDM, CoP/MoP1 *shall* 'elaborate on modalities and procedures with the objective of ensuring transparency, efficiency and accountability through independent auditing and verification of project activities' (Art. 12.7).

At present, it is still unclear what the executive board of the CDM will look like. Article 12 is quite specific about the role of CoP/MoP, and the position of the operational entities also seems to be quite clear. The role of the executive board, however, seems to depend on the extent to which CoP/MoP delegates activities to lower hierarchical levels. For example, CoP/MoP could play a relatively passive role and establish the executive board as a permanently active institution that continuously checks and verifies CDM activities. In this case, the executive board could act as a multilateral body that facilitates project development (for example via brokering or via a multilateral investment fund) and carries out verification and certification in cooperation with the operational entities. As such, the executive board would have strong similarities to a global or regional facility (for example, similar to the GEF), specializing in greenhouse gas emission reduction projects.

Another option is that CoP/MoP envisages a more active role for the level of operational entities. In that case, projects will be mainly established through bilateral cooperation between parties with the assistance of private or public operational entities. The role of the executive board would then be more passive and contain a supervision on the basis of 'best professional judgement'. Through an extensive report, the executive board could inform the CoP/MoP annually about the CDM activities and provide suggestions and recommendations.

An example of such a framework for the CDM could be found in the above-mentioned Costa Rican Protected Areas Project (Box 3.1).[9] This project recently issued so-called 'certified tradable offsets', (CTOs), which were achieved via a carbon sequestration project in Costa Rica. The project has been established by a group of international consultants at the request of the Costa Rican government. The CTOs were verified by a private institute specializing in the monitoring and verification of several types of projects. Finally, the whole project is supervised by a 'Certification Council', the members of which are international experts, including scientists, NGO representatives, policy makers and a representative of the IPCC.

A Proposed Structure for the CDM

On the basis of the above, a tentative institutional structure for the CDM could be proposed. Although the following description is not meant to be a

9. Foundation JIN (1998c).

blueprint for the final CDM structure, it may serve as a framework for further discussions.

The first level of the proposed structure is that of the project. An Annex I party investor (either a private sector party or a government) develops in cooperation with a non-Annex I party host a project that fulfils the basic requirements under Article 12 and the guidelines decided on by CoP(MoP). If the investing Annex I party government and the host non-Annex I party government agree on the project, they sign a letter of endorsement as an official approval. The monitoring of the project could be done by the project parties themselves, as it is precisely in their own interest that the project be successful and therefore carefully monitored. In the letter of endorsement, the parties would have to agree on the sharing of the proceeds of the project, on the extent to which certified emission reductions will eventually accrue to the investing party, and to on the extent to which the project contributes to the non-Annex I party's sustainable development.

Subsequently, the project developers will have to announce the project to the CDM executive board. They could do so by submitting a report similar to the Uniform Reporting Format under the AIJ pilot phase. At this stage, however, it does not seem to be necessary to have the project certified, although it is possible that project developers might wish to obtain some 'advice' from the executive board about whether or not the project can be eligible for ex post certification. Project developers that do not feel certain about their project may do this, although the extent to which they will ask for 'advice' will probably decrease as they become more experienced in developing CDM projects. On the basis of such a procedure, and assuming that CoP will give guidance on design options and guidelines for the CDM, projects could start as early as 2000. If CoP is able to provide sufficient clarity to project developers on how to establish CDM projects, they do not necessarily have to wait for the decision by CoP/MoP1 on who will certify the projects. This, however, depends to a large extent on the decision on how the baseline for CDM projects will have to be determined (see below).

Current CDM projects may be listed and published on a continuous basis (for example, via the Internet), so that the market for the CDM can be as transparent as possible. It could even be considered that planned projects, will have to be reported to the executive board in order to be eligible for certification as a CDM. This could prevent situations in which companies that invest in non-Annex I parties for commercial reasons, but that fail to make the investment profitable, apply for certified emissions reductions under the CDM afterwards. Although the project may have resulted in greenhouse gas emission reductions, it is not additional, as it was implemented for commercial reasons.

Another reason why the project may have to be announced to the executive board is the baseline determination. If CoP decides that the baseline for a project has to be approved beforehand, irrespective of whether it will be a project-specific, top-down or matrix baseline, detailed projections of the project's emission reduction contribution must be submitted for approval. In that case it may be doubted that the CDM can start in 2000, as the necessary institutions for approving baselines for the CDM may not be ready by then (see the second section of this chapter). In the case where CoP decides on an ex post baseline approval for the CDM, investors could start earlier on the basis of their own baseline (which, again, could be of a project-specific, top-down or matrix type).

Once the project is under way, the certification phase begins. When a project has ended, or when the project parties would like to receive their credits, they can submit a project report to the executive board for verification and certification. The executive board may subsequently pass the report on to the operational entities. The latter may be internationally recognized certifying institutions that have a long experience in verifying environmentally sound projects or investments. The operational entities may function in a number of different ways.

One option is to define a set of project categories, such as fuel switch, cogeneration, demand-side management, forestry (if allowed: see the previous section) and fugitive gas capture, and to appoint one certifying entity for each category. After the certification this entity submits a technical report to the executive board in which it says to what extent the project has contributed to the objectives of the CDM and how many certified emission reductions the parties may acquire. During this procedure the certifying operational entity would have to request the advice of an external review panel for each project category. Such a panel could meet every now and then, but each panel member could also give its review, for example, via E-mail. It would not even be necessary to let all members of the panel review the project's results. For each project, only two or three members could be requested randomly, as often happens with, for example, review articles for scientific magazines.

Another option is that for each category a number of certifying entities will be appointed who will be requested, for example, randomly, to certify CDM projects. After an entity has submitted its technical report, the executive board could start the review procedure itself, along the lines described earlier. In this case, however, the CDM executive board will have a more active role, as it will eventually decide, on the basis of the technical report of the operational entities and the external review, how many certified emission reductions each party gets.

Finally, an additional remark should be made on the composition of the executive board. As the CDM aims at both sustainable development for the non-Annex I parties and certified emission reductions that could assist Annex I parties in achieving their QELRCs, the executive board should have a balanced membership of representatives from Annex I and non-Annex I parties. In this respect the executive board may come to look like the council of the GEF and the Executive Committee of the Montreal Protocol, where the composition of the board ensures an equal say for all parties.[10] A similar remark should be made with respect to the review panels, the composition of which should also ensure that Annex I and non-Annex I parties are equally represented.

COVERING ADMINISTRATIVE EXPENSES AND ADAPTATION

In paragraph 8 of Article 12, the Kyoto Protocol states that CoP/MoP shall ensure that a share of the proceeds from CDM projects 'is used to cover administrative expenses as well as to assist developing country Parties that are particularly vulnerable to the adverse effects of climate change to meet the costs of adaptation'. By including this paragraph the Protocol, first, has 'covered' the issue of who will pay for the activities of the CDM executive board and, possibly, the operational entities. Second, Article 12.8 adds to the Protocol the issue of developing countries' adaptation to the adverse effects of climate change.

The issue of covering the administrative expenses can be divided into three basic questions: what is exactly meant by 'proceeds', what share of the proceeds will have to be used to cover administrative expenses, and who will cover these expenses? The answer to the first question may not be easy to give, as Article 12 is not very specific about the term 'proceeds.' Instead, Article 12.8 refers to 'proceeds from certified project activities', which could literally speaking, be cost savings for Annex I parties achieved by investing in a CDM project. It could also refer to economic benefits arising from the CDM project. Finally, it could refer to the value of the certified emission reductions in the 'market'.

With respect to the second question, the answer may depend on the eventual role of the CDM executive board (see earlier in this chapter). If the executive board's activities will be rather limited, its expenses may be lower.

10. See also Goldberg (1998, pp.21–22).

The real administrative expenses will, in that case, be incurred by the operational entities, which could invoice the project developers directly. If, on the other hand, the executive board will be a more active, operational body, its expenses will be higher, so that the project parties may have to remit part of their proceeds to cover the board's costs.

The third question of who will cover the administrative expenses – the investing party or the host party – is still an open one. It could be argued, on the one hand, that only the investor would have to pay the executive board on a project-by-project basis. On the other hand, it could be argued that both the host and the investor have to pay, as they both gain from the project, but that the investor should pay somewhat more than the host: for example, 75 per cent investor and 25 per cent host. Yet another possibility is that the investing countries pay for a subscription to the CDM executive board. In the latter case, Annex I parties that want to participate in CDM projects would have to remit a subscription fee to the executive board, from which the board's general administrative expenses could be paid. In this system, there does not have to be a direct link between a project and the subscription fee an Annex I Party has to pay. By paying a fee to the executive board, the Annex I party obtains a 'right' to begin a CDM project and to have it certified by the executive board.

The issue of assisting developing country parties that are particularly vulnerable to the adverse effects of climate change could be a difficult one. First of all, the three questions mentioned with respect to covering the executive board's administrative expenses hold for meeting the costs of adaptation as well. In addition, the issue is complicated by the fact that parties have to put aside funding, the final purpose of which is still rather vague. After all, if vulnerable developing country parties can become eligible for compensation in the case of damage due to climate change, it first has to be defined what (and whether) damage is caused by climate change. This is not an easy task. The IPCC Second Assessment Report (SAR) concluded that human activities have a discernible impact on climate change, but this conclusion was surrounded by several uncertainties. In fact, although it may be generally believed that the world takes a big risk if the greenhouse issue is not addressed properly, a clear relationship between human behaviour and climate change has still not been indisputably proven.

This complicates the determination of damage caused by climate change. The IPCC SAR has produced a considerable analysis of the damage that may occur in the case of human-induced climate change. However, it is not by definition possible to say that, if such damage occurs, it is caused by global climate change. In this case, the same argument that has been applied in climate change policies so far may hold. Although the relation between climate change and human behaviour is difficult to determine, it is better to start with policy now because, by the time the relationship is indisputable, it may be too late. One could therefore choose a system in which vulnerable developing countries can obtain funding to protect themselves against damage which is very likely caused by a human-induced change in the climate.

It may be clear that, in that case, clear criteria should be set in order to prevent abuse of the adaptation fund under the CDM. Possible criteria include the following.

- The level of GDP/capita of the country, which could be an indicator of vulnerability, as countries with higher income levels may be expected to be more able to protect themselves against climate change.
- The kind of damage which should be covered. Should the adaptation fund cover damage to agricultural harvests or should it be limited to activities such as improving the infrastructure in case a sudden evacuation of the threatened population has to take place? Or should the adaptation fund only be used for preventive investment in the developing countries concerned, such as improving sea dikes?
- What will be the price of the adaptation measure? Strictly speaking, the adaptation fund will function as an insurance policy for developing countries and will, thus, fill a gap that exists in the private insurance market, in order to be able to determine what amount of money will have to be put aside, for adaptation depends on the price to be paid for it. As reference/shadow prices may be difficult to find in the private insurance market (best professional judgement) estimates may have to be made for this.

THE CDM AND COMPATIBILITY WITH OTHER FLEXIBILITY MECHANISMS

Under the Kyoto Protocol, CDM may have a competitive advantage over JI because (CDM) certified emission reductions (CERs) achieved between 2000 and 2008 can be banked and (JI) emission reduction units (ERU) cannot. This could create a bias for parties towards postponing JI projects until shortly before 2008 (see also ETC/JIN, 1998, p.4).

Another possible competitive advantage of CDM projects over JI is the absence of emission targets in developing countries. Annex I parties that host JI projects have QELRCs, so they must deliberate whether selling the ERUs to the investing party reduces their potential to fulfil their commitments or creates a leverage to produce a larger amount of ERUs, which would not have been produced in the absence of the project (ibid., p.2). As long as non-Annex I parties do not adopt emission targets, they will not have to make this consideration.

As was discussed in the third section of this chapter, JI may have a competitive advantage over CDM, because sinks have been explicitly included in Article 6 (JI) and not in Article 12 (CDM). The possible exclusion of carbon sequestration through forestry projects for CDM would considerably deteriorate the availability of emission reduction possibilities in developing countries. Moreover, the adaptation and administration tax burdens the CDM only.

If CoP decides to remove the banking differences between JI and CDM, JI's competitiveness vis-à-vis the CDM would probably increase. In the case where banking of early emission reductions via CDM was deleted, however, the flexibility of both JI and CDM would be reduced. On the one hand, this might have a negative effect on the cost-effectiveness of such projects. On the other hand, developed countries would have more incentive to invest in emission reductions at home, which would correspond to their support for domestic action to reduce emissions. Finally, potential investing parties would have less incentive to invest in early emission reductions via CDM projects, which would not be supportive of developing countries' sustainable development policies.

To remove the differences with respect to forestry projects between JI and CDM, CoP could decide to introduce sinks in CDM Article 12. If sinks are introduced for CDM projects, the competitiveness of CDM would increase because developing countries have a large forestry potential, although the transaction costs for those projects could be higher than in central and eastern Europe.[11] A summary of the competitive advantages of the flexible instrument is given in Table 3.1.

11. The extent to which transaction costs for CDM sinks projects would be higher than for JI sink projects depends also on the host country. A non-Annex I party, like Costa Rica, has been very active in the AIJ pilot phase with forestry projects, so that considerable experience has already been gained, which may facilitate the implementation of future projects, thereby lowering the transaction costs.

Table 3.1 Comparing JI and CDM

Competitive advantage	Transaction costs	Banking of early emission reductions	Sinks
High	JI	CDM	JI
Low	CDM	JI	CDM

REFERENCES

CCAP/SEVEn (1997), *Joint Implementation Projects in Central and Eastern Europe*, description of ongoing and new projects prepared by Center for Clean Air Policy in Cupertino with SEVEn, background study to Regional Conference on Joint Implementation: Countries in Transition, 17–19 April 1996, Prague.

Chatterjee, K. (ed.) (1997), *Activities Implemented Jointly to Mitigate Climate Change*, Proceedings of the AIJ Conference held in New Delhi, 8–10 January 1997, Development Alternatives, India.

Chomitz, K.M. (1998),'Baselines for Greenhouse Gas Reductions: Problems, Precedents, Solutions', paper prepared for Carbon Offsets Unit, World Bank.

Dudek, D.J. and J.B. Wiener (1996), 'Joint Implementation and Transaction Costs Under the Climate Change Convention', Restricted Discussion Document ENV/EPOC/GEEI(96)1, OECD Environment Directorate/Environment Policy Committee, Paris.

ETC/JIN (1998), 'Dealing with Carbon Credits after Kyoto', Background Paper Expert Meeting, ETC Consultants in Development Programmes (ETC) and Joint Implementation Network (JIN), 28–29 May, Callantsoog, The Netherlands.

Foundation JIN (1997a), 'Baseline Determination: Summary of Thoughts', *Joint Implementation Quarterly,* 3(3), 7.

Foundation JIN (1997b), 'Nordic Study on 10 Projects in Eastern Europe', *Joint Implementation Quarterly* JIQ 3(4), 10–11.

Foundation JIN (1998a), 'Costa Rican Carbon Offsets Certified,' *Joint Implementation Quarterly,* 4(2), Renewable Energy-Grid Project in Mexico, approved by USIJI, 10–11.

Foundation JIN (1998b), *Joint Implementation Quarterly*, 4(2), 11–12.

Foundation JIN (1998c), 'Planned and Ongoing AIJ Pilot Projects', *Joint Implementation Quarterly,* 4(1), 14.

Goldberg, D. (1998), *Carbon Conservation: Climate Change, Forests and the CDM,* CIEL, CEDARENA, Washington, DC/San José.

Harmelen, A.K. van, S.N.M. van Rooijen, C.J. Jepma and W. van der Gaast (1997), *Joint Implementation met Midden- en Oost-Europa*, ECN/JIN (in Dutch), Groningen.

Karani, P. (1997), *Constraints for Activities Implemented Jointly (AIJ) Technology Transfer in Africa*, Washington.

Michaelowa, A. (1998), 'Joint Implementation – The Baseline', Global Environmental Change, 8(1), 81–92.

Nordic Council of Ministers (1997), 'Criteria and Perspectives for Joint Implementation', *TemaNord*, Copenhagen, 51–7.

4. CDM and its Implications for Developing Countries

Sujata Gupta and Preety Bhandari

CDM DEBATE

The CDM has been proposed in the Kyoto Protocol with the twin objectives of helping Annex I countries achieve their emissions reduction targets and helping non-Annex I countries promote sustainable development in their economies. In addition, funds for CDM projects are to contribute a small (so far unspecified) percentage towards an adaptation fund to help vulnerable countries combat or else adapt to climate change. The Kyoto Protocol states that the three mechanisms specified, namely, CDM, joint implementation (JI) and emissions trading (ET) have to be supplemental to domestic action towards meeting the quantified emission limitation or reduction commitments (QELRC) of the Kyoto Protocol. The limits for the three mechanisms or for individual mechanisms have not been specified in the Kyoto Protocol.

In a review of the likely demand and supply of emission reduction units (ERU), it has been estimated that demand for carbon abatement that will originate in OECD countries in 2010 will be about 1 billion tonnes of carbon (Figueres, 1998). The projected availability from economies in transition is about 165 million tonnes of carbon, which is essentially attributable to 'hot air'. On the other hand, the potential from JI is about 250 million tons of carbon. The balance is expected to be met by domestic action within Annex B countries as also CDM projects. The prospects for CDM activities, prima facie, are bright in light of the market for credits and also by the early start bestowed upon it by the Kyoto Protocol. However, the relative ease of acquiring credits from hot air, and the possibility of a relatively higher cost of certified emission reductions (CERs) from CDM,[1] could detract from the use of this mechanism. This may be compounded by the fact that an early start

1. CDM costs are higher as a result of the imposition of administrative charges and a contribution towards the adaptation fund.

for the CDM entails experimentation and associated uncertainties that will be resolved for mechanisms to be implemented at a later stage.

It was made clear at CoP4 that attention would be focused on CDM, as this mechanism is likely to be the first to be operationalized. It is expected that by 2000 a minimum set of modalities for CDM will be agreed to enable the launching of CDM projects, at least on an experimental basis. This chapter discusses issues related to CDM and a possible structure for CDM incorporating the guidelines of the Kyoto Protocol. The last section discusses how the concerns related to CDM can be addressed.

POLITICS

The debate on CDM has a broad range. Some developing countries have concerns in adopting the CDM as it will enable Annex B countries to fulfil all their emission reduction targets through this instrument without undertaking any domestic action. Annex B countries, on the other hand, argue that, given their agreement to control their emissions, they should be allowed to select the least expensive option, in the interest of efficiency. In fact some have argued that insistence on supplementarity will raise the total costs of abatement owing to higher costs of domestic action in Annex B countries. Others make claims that supplementarity will not benefit the developing countries, as it will constrain the demand for offsets likely to be generated in non-Annex B countries.

The arguments essentially mean that, while most developing countries are favouring a cap or a limit on the extent of emission reduction that can be achieved using the CDM alternative, Annex B countries, with some exceptions, are for no limits on recourse to the flexible mechanisms.

Another concern of the developing countries is regarding giving away economical options of reducing emissions at throwaway prices to Annex B countries under CDM projects. There is apprehension that the 'low-hanging fruit' or the less expensive mitigation options will not be available at a later stage when the host countries could be required to reduce greenhouse gas emissions and that this in turn would entail undertaking abatement at astronomical costs.

INSTITUTIONAL ASPECTS OF THE CDM

Recent literature and debates on CDM have highlighted the various aspects that need to be resolved. These can be categorized as those relating to governance, baselines, monitoring and verification of offsets. The governance aspect mainly relates to the institutions – national and international – that will be a party to the functioning of the mechanism. It begins with the developing country institutions that will approve the projects, international institutions or the executive board which will act as a regulator responsible for examining and verifying projects, designating independent auditors for monitoring and certification of credits and resolving any disputes that may arise.

Within the ambit of overall governance is the crucial aspect of developing baselines. There is a plethora of literature on the various types of baselines (Puhl *et al.*, 1998; Heister, 1998; Baumert, 1998) – static versus dynamic, top-down versus bottom-up – the merits and demerits of different types of baselines. Added to this is the question of the high transaction costs that may ensue and the necessary expertise that will be required for determining the baselines. Furthermore, there are concerns about the lack of data required to determine baselines for projects in developing economies.

The task of establishing baselines still looms large in any discussion on CDM. The contentions about national versus project, static versus dynamic baselines notwithstanding, it has to be decided whether this task would be vested with an international or a national body. There are obviously advantages in having an international body, as long as it has all-round acceptability and especially from developing countries. This would relieve the burden on the already overstressed government departments in developing countries. In this context, the UNFCCC secretariat seems one such option; the secretariat could undertake this with the involvement of government designated NGOs or relevant technical consultants from within the host country, to give it more credibility.

Critical to the consummation of CDM projects are the carbon offsets that will be generated. A conundrum that may arise with respect to CERs is the case where multilateral financing for projects is involved. This could be compounded by the fact that involvement of multilaterals provides leverage for bilateral financing for a project, leading to a multiple ownership of CERs. The first issue that needs to be resolved is allocation of CERs to multilateral bodies such as the World Bank. Would this be permitted? If so what would be the rules of the game for these multilaterals to sell these CERs in a secondary market. Several questions need to be answered. Who will monitor these? Who will certify them? When will they be credited – at the beginning of the project or at regular intervals? What will be the sharing mechanisms for the carbon offsets? Will a secondary market evolve? The answers to the above questions will depend to a large extent on the framework that will be designed for CDM and the market for CERs.

This chapter suggests a model for the CDM that is flexible (given the range of issues that need to be addressed) and works within the guidelines contained in the Kyoto Protocol. As time is of the essence, it is essential that a skeletal framework for CDM be elaborated and tested.

STRUCTURE FOR IMPLEMENTING THE CDM

Michaelowa and Dutschke (1998) discuss three possible frameworks, namely a multilateral fund, a clearing house and a project exchange. Several studies have discussed a multilateral and bilateral framework. The broad features of these forms are discussed below.

In a multilateral fund, investments are channelled through the fund and all credits accrue to the fund. The fund is controlled by the executive board and all projects are registered with it. The investors are protected from risks of non-materialization of credits. Investors can participate with a small amount of capital. There is no one-to-one link in the investment and carbon credits accruing from projects. This could be a framework that is favoured by small investors and small host countries.

In a bilateral approach, the CDM project is viewed primarily as a transaction between two entities. Either or both of them could be government or private parties and they reach a mutual agreement on a specific project: its financing, sharing of the likely carbon credits and risks. Each bilateral agreement would be governed by the rules and regulations of the executive board regarding the additionality criteria, verification, monitoring and certification processes.

The CDM is a project-based instrument and a structure based on a bilateral approach is discussed here. The approach can be altered to take on a multilateral basis. The main difference in the two approaches is that, in a multilateral framework, the fund operates as a clearing house for CDM projects, while in a bilateral model this function is replaced by the market.

The detailed functioning of the proposed model is discussed below (see also Figure 4.1). The proposed structure for CDM is desirable, as it is amenable to all permutations and combinations that may arise in launching CDM projects.

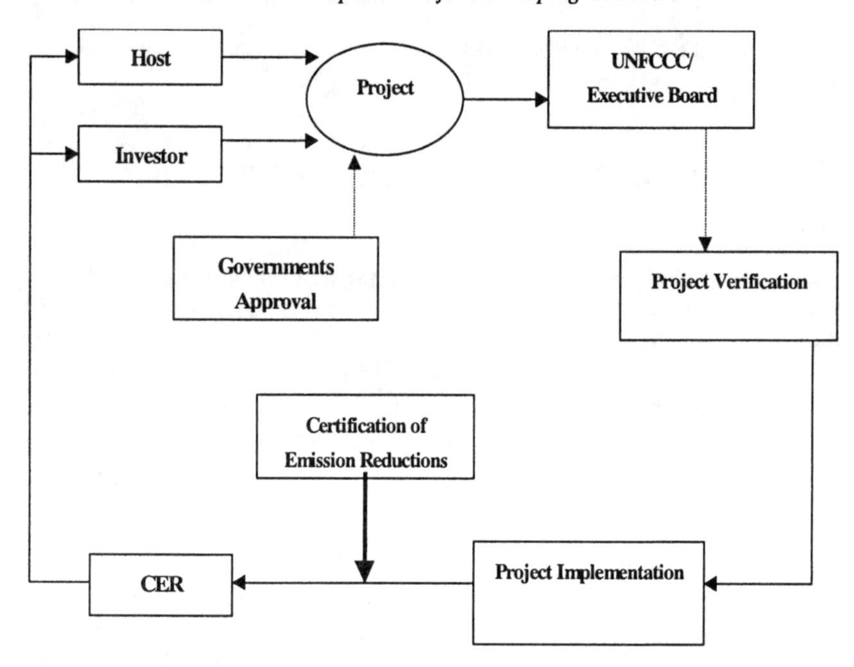

Figure 4.1 Institutional structure of the CDM

- The host and investor together identify a project that qualifies as a CDM proposal.
- Necessary government approvals are obtained by both the investor and the host party.
- The project is verified by the UNFCCC secretariat or the executive board or an independent agency, and clearance is granted with respect to the project meeting the financial and environmental additionality criteria. This would entail the determination of a baseline for the project.

Once the above issues are cleared, the following action is taken.

- The host and investor reach an agreement on the terms regarding sharing of emission reduction credits, the shares in the cost of the project and issues related to non-compliance and default.
- The project is implemented.
- A contribution is made towards the adaptation fund and the administrative costs. This is viewed by us as a tax on CDM vis-à-vis the other Kyoto mechanisms, namely joint implementation and

emissions trading. The contribution to the adaptation fund could be made when emission reductions are certified.[2]

- Emission reductions from the project are certified as and when they happen based on an agreed baseline, and CERs are created with the implementation and operation of the project

Some papers have argued that CERs can be challenged at a later date (Werksman, 1998) on the basis that a wrong methodology was used or a project was destroyed. This is not necessary if CERs originate when the reductions have actually happened and were certified. At the present time, sequestration projects are not included under CDM projects. In cases where sequestration projects qualify as CDM projects, the executive board can decide on how to deal specifically with these types of projects. In case where CERs generated are challenged on account of a wrong methodology it is suggested that the following process will resolve the issue and, in addition, assist in addressing a few of the baseline issues.

In setting rules and guidelines for the verifying and certification agencies, the executive board can indicate that (in the interest of dynamic baselines) after a specific period, say five years of operation of the project, the baseline could be revised, keeping in mind technological developments; however, the baseline will not change more than a specified percentage over the preceding period's baseline for projects that are already sanctioned. Further different time periods and ranges for percentage change can be identified for different categories of projects. For example, a power project based on gas having an efficiency of 55 per cent is compared with a coal-based power plant with an efficiency of 35 per cent as the initial baseline. After a period of five years, the average efficiency of coal-based power plants in the country may increase to 38 per cent, which then becomes the new baseline for the existing project, provided that change in the new baseline does not exceed 20 per cent. Such a system provides a limit on the extent of uncertainty and does not prevent efficiency improvements in existing or sanctioned projects.

ROLE OF STAKEHOLDERS

There are four different stakeholders in the model suggested above: the executive board, the governments of the host and the investor parties, the investor and the partner entity in the host country.

2. It is proposed that a similar contribution be required from all the Kyoto mechanisms.

Executive Board

The executive board plays the role of a regulator. Its main task is to ensure that the carbon savings from the project would not have resulted in the absence of the project and hence it meets the environmental additionality criterion. In order to establish environmental additionality of the project it becomes necessary to determine the baseline. Therefore the executive board will have to evolve criteria to determine baselines. Once the project is verified and implemented, the board has to designate independent monitoring and certification agencies for each project. All certified emission reduction units generated will be registered with the executive board.

The tasks of the executive board can be classified as follows:

- establishment of baseline criteria,
- verification and establishment of environmental additionality (through an independent agency),
- designating a monitoring and certification agency,
- registration of CERs.

Governments

The other major stakeholders are the governments of the investor and the host party. Each party has to gain the approval of their government. The investing country government should have clear guidelines on the reservations it may have in executing CDM projects with certain developing countries or for certain type of projects (for example, nuclear projects). The provision for seeking approval for CDM projects will assist the government in monitoring the quantum of carbon reductions likely to be achieved through these projects. This would be particularly important in cases where there are stringent caps or limits on the extent that can be achieved through the flexible mechanisms. In addition, where the investor country (an Annex B party) may have allocated targets or reductions to be achieved through the CDM across different sectors in the economy, the process of seeking approvals will facilitate this process as well.

In the host country, as a first step, the government needs to prioritize the type of projects it perceives as potential CDM projects. These projects should meet the financial additionality criterion and should make a contribution to development in the economy. The concern that developing country governments have that CDM projects will trade away the less expensive abatement options should be addressed at this stage. There can be several projects that could qualify as CDM projects in meeting the environment and financial additionality criteria. The government can identify criteria other than carbon abatement to evaluate the project. Table 4.1 lists four hypothetical projects and additional parameters that the project must address. The host country may decide to exclude projects of type A as it fears that it is trading 'low-hanging fruits' and the additional benefits that result are not significant. At the other extreme, type D projects with high carbon abatement costs and favourable benefits could clearly be included in the list of desirable CDM projects. It is for the government to decide whether to include projects of type B and C. The government should clearly prioritize projects and define rules of exclusion for potential CDM projects. Once the preferred projects are determined, this information should be in the public domain.

Table 4.1 Evaluation of project benefits by host countries

Project	Cost of carbon abatement $/ton of carbon	Employment generation	Positive local environmental impacts	Access to state-of-the-art technology
A	5	Low	Low	Low
B	10	High	Medium	Low
C	30	Low	Low	High
D	50	Medium	High	High

The host country government may want to impose a tax on the transfer of carbon credits through CDM projects. Again these decisions have to be clearly communicated so that the investors are aware of the different costs related to the project. The host country government has to take into consideration, if it excludes the less expensive CDM projects and imposes any additional costs in the form of taxes, CDM projects from the country will be put at a disadvantage in competing with CDM projects from other countries.

The tasks for concerned governments, of both host and investor countries are:

- to prioritize the type of projects they are willing to consider as potential CDM projects,
- to clearly define rules of exclusion,
- to provide easy access to the above information,
- to specify the tax structure, whether monetary or in terms of carbon credits.

Project Entities

The other two major stakeholders in a CDM project are the investor and the host country party participating in the project. As these are entities undertaking investments, they have to ascertain that each party has obtained all necessary approvals and clearances from the concerned governments and the executive board. Next the terms of investment have to be settled, that is, whether the investment will be equity participation or whether it will be a loan from the investor to the host country party. In the case of equity participation, are the rights limited to carbon credits produced by the project or do they extend to other goods (say, electricity-generated) produced by the project. Further, they have to include in the contract the terms for sharing the carbon credits and addressing issues related to non-compliance of the project.

The issues the investor and host party will have to address are:

- clearances and approvals from governments and the executive board,
- the form of financial collaboration,
- rules for sharing of credits,
- risk sharing in the case of non-compliance of carbon offsets.

CONCLUSIONS

There is no denying the fact that CDM provides developing countries with an opportunity to attract much-required financial flows. The ways and means a country adopts to elicit such funding, however, need to be clearly chalked out. Transparency in the project selection and approval process is of utmost importance. The foremost criterion for the host country is to undertake a national exercise in prioritizing the sectors where such funding or projects should be undertaken, which should be in line, obviously, with the development objectives of the country. Such an exercise itself is a gigantic task, especially in light of the political wranglings and multiplicity of controls normally prevalent in a developing country. Once a relative ranking of sectors is undertaken, projects can be identified within these and a menu of project ideas can be prepared. An international body like the UNFCCC secretariat could be entrusted to keep a roster of these projects. The roster (posted on an Internet website) could be accessed by interested bilateral, multilateral or private investors from Annex B countries or even private parties within the host country, for further development of projects. In fact involvement of the private sector in the host country has the dual advantage of capitalizing on their information base on feasible technological avenues and synchronizing with the current move in some developing countries towards privatization.

However, the key issues that need to be addressed simultaneously relate to setting up clear guidelines on who the approving authority would be in the host country, and which principles for approval it would use.

Involvement of international bodies in monitoring and verification as well as certification is important, as developing country institutions may not possess the wherewithal to undertake this function. Further, rather than developing such multifarious institutions in all developing countries, it will be economical to have a one-point responsibility for this function.

Another issue that deserves attention especially from the developing country perspective is the generation of carbon offsets from projects and the host country's share in the same. It is prudent to argue that a share of these offsets should accrue to the host country, but the sharing mechanisms and rules for the same will need to be devised at the outset. This leads to another issue in the context of exclusive involvement of private parties in a CDM project: in some instances, the host government may ask for a share of credits, perhaps in the form of a tax. International guidelines will have to consider this aspect of credit sharing in ensuring that there are no rules that preclude such a taxation regime for CERs generated.

As identified in this chapter, a simple institutional structure is required for the functioning of the CDM. Outlining of this structure basically entails the parties being directly involved in the projects, the institutions that will be external to the project and facilitate its functioning. The debate on bilateral and multilateral frameworks is counterproductive, in that there should be one overall framework within which all possible project ideas can be dealt with. Even though the structure is simple, the issues, options and arguments that may arise at each step need to be examined at the outset. The main contentious issues, we feel, would relate to baselines and CERs.

Many observers perceive that developing countries will project inflated baselines, deliberately pursuing business-as-usual development patterns. This fear has to be jettisoned. Also any disagreements over CERs, their ownership and sharing need to be avoided. Finally, the preparedness of developing countries in undertaking CDM is essential to kick-start the process.

Therefore, an experimental phase should be launched, with a shared learning for all concerned parties, as also for institution building. The fear that this experimental phase may meet with a fate similar to that of the pilot phase of AIJ is not relevant, as in the CDM there are tangible benefits in the form of CERs. Central to this is the need to ensure a mutual and participatory approach to global solutions to mitigate climate change.

REFERENCES

Baumert, K.A. (1998), 'The Clean Development Mechanism: Understanding Additionality', The Clean Development Mechanism Draft Working Papers, October, World Resources Instititute, Foundation for International Environmental Law and Development and Center for Sustainable Development in the Americas, Washington.

Figueres, C. (1998), 'How Many Tons? Potential Flows through the Clean Development Mechanism', The Clean Development Mechanism Draft Working Papers, October, World Resources Instititute, Foundation for International Environmental Law and Development and Center for Sustainable Development in the Americas, Washington.

Heister, J. (1998), 'Certification of Baselines and Verification of GHG Offsets', draft paper, Carbon Offset Unit, World Bank, Washington.

Michaelowa, A. and M. Dutschke (1998), 'Creation and sharing of credits through the Clean Development Mechanism under the Kyoto Protocol', paper presented at the experts workshop, 'Dealing with carbon credits after Kyoto', Callantsoog, The Netherlands, 28–9 May.

Puhl I., T. Hargrave and N. Helme (1998), 'Options for Simplifying Baseline Setting for Clean Development Mechanism', Center for Clean Air Policy.

Werksman J. (1998), 'Compliance issues under the Kyoto Protocol's Clean Development Mechanism', The Clean Development Mechanism Draft Working Papers, October, World Resources Instititute, Foundation for International Environmental Law and Development and Center for Sustainable Development in the Americas, Washington.

5. Renewable Energy Supply Systems in Indonesia: a Case Study

Regina Betz

INTRODUCTION

The Renewable Energy Supply Systems (RESS) project is an AIJ project carried out by the E7 initiative combining eight of the world's largest electric utilities of G7 countries (RWE AG Germany is the main coordinator of this project). The RESS project's purpose is to supply a limited but reliable amount of electricity to households and community facilities in remote areas in Indonesia by harnessing solar, wind and hydro energies. Consequently, living conditions could be improved and GHG emissions simultaneously reduced.

The idea seems to be really good, but how good is the implementation? The present chapter aims to answer this question and to demonstrate problems within the AIJ pilot phase so as to provide information based on practical experience for future project-oriented mechanisms such as joint implementation (JI) or the clean development mechanism (CDM). The first section of the chapter gives a description of the project's background, organization and different components including the financial aspects and the present state of implementation of the RESS project. The second section presents the evaluation of the project: primary objectives are analysed from the investor's view and secondary objectives more from a development cooperation and political point of view. In the final section general problems and especially problems relevant to the further development of the CDM are summarized.

INDONESIA

Indonesia consists of 13 600 islands and the impact of global warming would severely affect its economy (GTZ and BPPT, 1995, pp.99). Therefore the Indonesian government ratified the UNFCCC in 1994. It, for instance, has initiated two greenhouse gas mitigation strategy studies clarifying the relevance of renewable energy technologies as a greenhouse gas abatement option. This is particularly important for Indonesia in view of the significant potential these technologies have in the country. In addition, Indonesia may be forced to import fuels to satisfy the growing demand for energy in the long run. The government intends to meet this challenge with measures for increasing energy efficiency and diversifying energy resources (Bundesstelle für Außenhandelsinformationen, 1994, p.5). Fostering the application of renewable energy plays a significant role in this context. Finally, the Indonesian government has supported the AIJ pilot phase by signing several letters of intent with potential investors. One of these AIJ projects will be analysed in the following.

PROJECT DESCRIPTION

Background

Only 21 per cent of total households in the province of Nusa Tenggara Timur (NTT) are currently connected to the power grid. Thus NTT so far is one of the least electrified regions in Indonesia. Changes in the near future will be unlikely because the grid connection of rural villages would incur prohibitively high costs for the state electric power company, Perusahaan Listrik Negara (PLN). The thin settlement of provinces outside Java may be the main reason for these high costs. Power supply nevertheless is a core element in improving the living conditions of the inhabitants of remote areas. Furthermore, it is listed as one of the primary objectives in the Second Long Term Strategic Development Plan (1994–2019) of the Indonesian government. In this plan, the electric power supply of all rural areas before the end of REPELITA VII (2003–4) is aimed at. However, according to evaluations by the Agency for the Assessment and Application of Technology (BPPT), 40 per cent of all households will still be without a power supply, since a village is regarded as electrified if only one household in the village is connected to the grid (BPPT, 1995, p.1).

Aim of the Project

The purpose of the project is to supply households and community facilities in remote areas in Indonesia with a limited but reliable source of renewable electricity and to gain practical experience during the AIJ pilot phase. The project should show how economically and socially self-sustaining facilities for electricity supply in remote areas can be established without generating additional greenhouse gas emissions. The project will also help investors to evaluate future business opportunities.

Project Proponent

The E7 initiative is coordinating and financing the project. The expert network was founded in 1992. The aim of the initiative is to cooperate and participate actively on global electricity-related issues, with an emphasis on the global environment: 'The common goal [...] is to "play an active role in protecting the global Environment and in promoting efficient generation and use of electricity"' (E7, 1997, p.5).

Today the E7 includes eight large power supply industries from six G7 countries, which at present cover about 11 per cent of world energy demand:

- Electricité de France (EDF) (France),
- ENEL (Italy),
- Hydro Quebec (HQ) (Canada),
- Kansai Electric Power (KANSAI) (Japan),
- Ontario Hydro (OH) (Canada),
- RWE AG (Germany),
- Edison International (EI) (USA),
- Tokyo's Electric Power Company (TEPCO) (Japan).

In January 1998, the experts network had about 29 active and 18 completed projects. Three of these projects were declared as AIJ projects, and one of these is the Renewable Energy Supply Systems (RESS) project in Indonesia, which will be examined in the following (Adam, 1999, pp.3–4).

Organization and Management

Several institutions are involved in the organization of the project. Figure 5.1 illustrates the project organization and indicates the most essential institutions and project components with their relation to each other.

The letter of intent (LOI) concerning the cooperation on activities implemented jointly (AIJ) is the basis for a memorandum of understanding (MOU) between the government of Indonesia and the E7. At the signing of the LOI and the MOU, the E7 was represented by John Fox (OH). In the MOU, the Directorate General of Electricity and Energy Development (DGEED) appointed the Director for Energy Development to act as the executing agency in cooperation with the Agency for the Assessment and

Application of Technology (BPPT) and the National Institute of Aeronautics and Space (LAPAN). Furthermore, thanks to the MOU the DGEED will contribute to approximately 1000 solar home systems (SHS) including CO_2 reduction monitoring, administrative support, tax exemptions for personal income of experts and so on. Perusahaan Listrik Negara (PLN) is the state enterprise for gas, oil, coal, geothermal and electricity supply and is supervised by the Ministry of Mines and Energy (MME).

The steering committee is the overall governing board for all E7 network projects. For the implementation of the RESS project a team of experts was established from the E7 including a project leader, a residence project manager and six task leaders. The project leader is Hans-Georg Adam of RWE, Germany. He is responsible, for instance, for the general implementation and management of the project and the internal/external communication and coordination within the project team. Claus Dauselt is the Resident Project Manager of the local project office which was established in early 1997 in Kupang (NTT). He runs this office, and is on the one hand responsible for procurement and accounting for goods and services within the project, as well as the development and implementation of local training concepts, seminars and so on. On the other hand, he has to give support to the coordination and cooperation with the Indonesian authorities involved, to select the plant sites, to survey the installation and to conduct the socioeconomic surveys. There are six different 'tasks' and 'task leaders', four of which will be examined in more detail in this paper: small photovoltaic (PV) systems, so-called 'solar home systems' (SHS), four micro hydro power plants (MHP), one hybrid system (HS) and the socioeconomic integration (SEI), including a Statistical Household Income Survey (SHIS).

The task leaders have to manage their tasks and prepare individual work plan schedules, and are responsible for the technical and specific results in their area (E7, 1997, p.12).

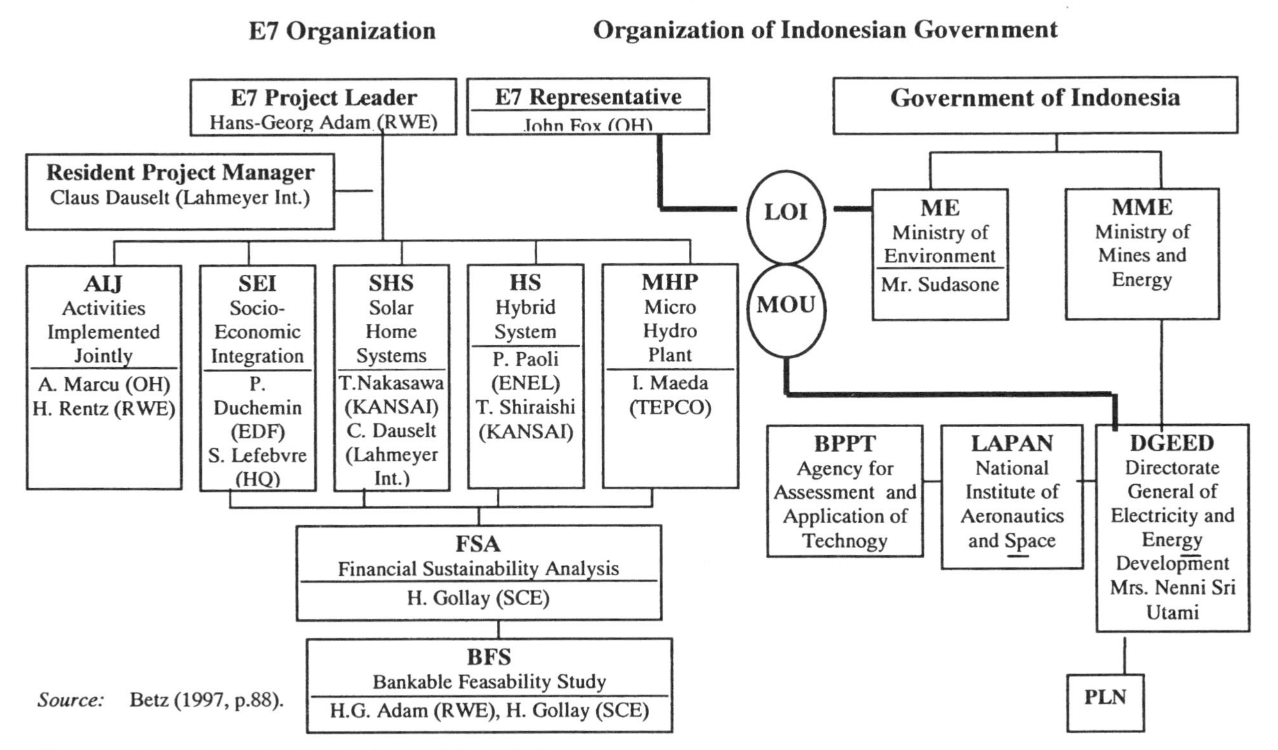

Source: Betz (1997, p.88).

Figure 5.1 Organizational chart of the RESS project

Project Tasks

Solar home systems (SHS)

According to the current state of information the 'actual number of presently 100 systems to be installed is likely to grow' (E7, 1998a, p.33). These 100 SHS were financed by the E7 as the Indonesian government is not able to meet its commitments under the MOU and to provide the requested 1000 SHS.[1] However, the DGEED provided 100 SHS and hopes to provide all the envisaged 1000 SHS within the coming fiscal years (E7, 1998, p.3).

Based on the Statistical Household Income Survey (SHIS), two villages in NTT have been earmarked for SHS implementation, selected from the seven surveyed villages (EDF, 1997, p.2). Lengkonamut on Flores is the best village for SHS. It has a relatively high income level (22 per cent of all households had an income of US$100–200 a month and 78 per cent less than US$ 100 a month) and a potential for regular agricultural development in the near future. Furthermore, its remote location in the mountain area would make a connection to the PLN grid most unlikely (too expensive).

Oinaineno on Timor has the advantage of homogeneous income sources (18 per cent of households had an income of US$100–200 a month and 82 per cent less than US$ 100 a month) and a potential for agricultural production development due to sales opportunities in Kupang. Above all, the geographical location prevents any connection to the PLN grid in the near future, and a road offers easy access from Kupang.

Taking into consideration that the SHS suppliers should be represented locally for after-sales and guarantee reasons, the local Indonesian company CV Jacobs was selected and awarded the procurement of the 100 SHS. The contract was signed in July 1998 and installation works were planned to start in October/November 1998 (E7, 1998a, p.3). The installation was finished by mid-February 1999.

In the standard case (see Figure 5.2), one 50Wp (watt photovoltaic power) SHS per household will be installed, which will provide the household with a 12V direct current. An SHS usually consists of a solar panel, a battery, a battery charge controller, an outlet for TV or radio, switches bulbs and/or fluorescent lighting. The reconstruction of the 18 SHS in Kualeu[2] has shown that it is very important for generating the maximum amount of electricity to

1. These SHS are to be provided by the Bavarian–Indonesian Joint Solar (BIGSOL). The Indonesian funding in this project would be about 50 per cent of the total US$40 million (Adam, 1997b, p.3).
2. The E7 decided as a spontaneous side activity to reconstruct 18 SHS in Kualeu which were installed in 1994 with government support. This reconstruction was completed in December 1997 and the E7 gained a lot of knowledge about the problems which can occur by introducing solar products in Nusa Tenggara Timur (E7, 1998a, p.9).

mount the solar cells properly at the right angle and in the right direction. Moreover, it is necessary to fix the wiring with cable staples in order to prevent loose wiring and to cover the battery using BCU housing boxes for safety reasons (E7, 1998a, p.10).

Source: Schweizer (1996, p.107).

Figure 5.2 Outline of a solar home system

Assuming 3.4 hours of sunshine per day, the 50Wp SHS will produce 170Wh/d. This generated electricity will enable a household to operate one 15W fluorescent lamp for five hours and a 6W bulb for 6 hours. The remaining electricity (59 Wh) can be used for the operation of a small black and white TV and/or a radio. Experience from some other photovoltaic-electrified villages showed that the inhabitants were using self-made 2W car bulbs to increase the light hours or number of lights (Betz, 1997, p.94).

The original plan of the E7 was to establish a self-financing concept (SFC) that would allow rural households to pay a monthly rate (the level to be determined by SHIS) to cover operating costs and to build up a reserve fund for future additional or substitute facilities (E7, 1997, p.25). Taking into account the results of the SHIS, the E7 decided to modify the concept and created the financial sustainability analysis (FSA) which mainly focuses on covering operation and maintenance (O&M) costs. As preliminary cash flow calculations have indicated, there will be sufficient revenues generated to pay for recurrent O&M costs of the SHS even with the prevailing poor economic conditions in Indonesia (E7, 1998a, p.32; 1998b, p.12).

The establishment of an effective organization to collect the fees and control the funds for O&M and manage spare parts is a main precondition for the FSA. The E7 had its first experiences in this area during the training sessions in Kualeu. The initial step was to set up a new village utility (Perusahaan Listrik Desa, PLD) and to replace the previous one established by the village unions (Koperasi Unit Desa, KUD), which had no specific instructions in the managing of SHS. The members of the PLD were selected by the SHS user group of Kualeu, and consist of a chairman, a vice-chairman (and secretary), a cashier, a person responsible for marketing and spare parts and four technicians. The personnel of the PLD were trained on utility performance procedures such as monthly reporting, bookkeeping and documentation, spare part management and customer service. The technicians were trained in the principles of electricity, the outline of the SHS, how to maintain an SHS, troubleshooting and in their special role as technicians. After the introductory training sessions, half of the technicians were not confident about undertaking maintenance activities. Further training on the planned troubleshooting simulator is necessary (E7, 1998a, p.11). At present (spring 1999), the PLD in Oinaineno has been established and the basic training has been carried out.

Micro hydro power plants (MHPPs)

As the Second Progress Report of the E7 stated, three micro hydro power plants (MHP) will be established and one MHP will be refurbished. TEPCO, which is responsible for this subtask, recommended that the E7 award a construction contract to the Japanese company KANDENKO. The contract period runs from 26 August 1998 to 29 October 1999 and comprises the construction of all four plants. See Table 5.1 and Figure 5A1.1 for sites that were selected for the MHPPs (E7, 1998a, p.14; 1998b, p.17):

KADENKO has to coordinate the work relating to material procurement and manage the construction work, as well as developing operation and maintenance manuals for the MHPPs. It is required to subcontract with an Indonesian company and to use Indonesian products if available. The land acquisition will be negotiated between Dauselt and the KUDs who procure the official written documents. The land will be used free of charge, but compensation for the loss of crops caused by the construction of a powerhouse, penstock and headrace is planned (E7, 1998a, p.14)

Table 5.1 Sites selected for the MHPPs in Indonesia

New MHPPs (South Sulawesi)	Power output	Numbers of customers in		
		2000	2001	2002+
Taba	46kW	176	282	352
Tendan-Dua	64kW	190	303	379
Bokin	13kW	52	83	104
Refurbishment (Sumba)				
Waikelosawa	15kW			

Note: All generated electricity will be sold to PLN.

The MHPP in Waikelosawa was built in 1976 with funds provided by West Sumba province and was struck and damaged twice by lightning. After being damaged the second time it was left unrepaired, the electrical equipment and metal works having to be replaced by new equipment (E7, 1997, p.18). The electricity generated by this plant will be fed into an existing grid and sold to the PLN, which means that a power purchase agreement (PPA) was required. After the PPA was signed by the PLN in February 1999, the refurbishment started immediately.

Hybrid system (HS)

The hybrid system for which ENEL is responsible will comprise a photovoltaic unit with a wind generator and a diesel back-up. Batteries will guarantee electricity supply for three days. If the solar power and wind generator are out of service a diesel generator will serve for emergencies. The electricity produced will on the one hand be delivered directly to households via a village grid, and on the other hand be used for battery charging for off-grid supplies. Most likely equipment will be supplied by Indonesian firms for guarantee reasons. Only inverters, rectifiers and electric boards are expected to be procured from Italian manufacturers who have shown reliable results in previous supplies to ENEL (E7, 1998a, p.18).

After several investigations of different sites, Oeledo on Roti Island was selected as the most suitable for the HS installation. Oeledo had a better economic structure, technical personnel were already available and there was a strong wish among the population to obtain electricity (ibid., p.27). In 2000, about 80 households and in 2001 the remaining 20 households will be connected to the grid of the HS. The design guidelines were oriented towards high operation reliability and minimum ordinary maintenance. The main components have been planned as listed below (E7, 1998a, pp.4, 18; 1998b p.5):

- *Photovoltaic panels* 256 crystalline silicon modules for a nominal power of approx. 21.6kW$_p$
- *Wind generator* Three blades wind-turbine with a nominal AC power of 10kW
- *Battery storage* 120 Pb-acid elements, for a rated C10 capacity of 144kWh
- *Inverters* Two 20kVA rated power self-commuting inverters (60kVA as peak capacity)
- *Diesel generator* Diesel generator with a nominal power of 20kVA as emergency reserve
- *Rectifier* A 10kVA rated power AC/DC converter for the energy from the diesel to the battery
- *Battery charging station* Three 12V output battery chargers
- *Circuit breakers* To protect the system

Up to now, a PLD has been established and the project and the basic elements of solar energy use and electricity have been explained to the village population, which is undergoing a basic training. The technical personnel will be trained during the installation phase and the trial operational period in order to learn ordinary maintenance, troubleshooting and simple repairing (E7, 1998a, p.21).

In the second progress report it was assumed that the procurement contract could be awarded by April 1999, so the on-site work could start after the rainy season in May for operation in August of the same year. The procurement plan and the placement of the major component orders was planned to be ready by the end of 1998. A detailed visit to the site with potential contractors for civil works and the distribution network was expected to follow in January 1999 (E7, 1998a, p.22).

Socioeconomic integration

The socioeconomic integration study examines the three subcomponents under the following criteria to ensure compatibility with the surrounding environment: (a) Environmental – insolation strength, direction and regularity of winds, water inflow, landslides and so on; (b) socioeconomical – acceptance of new technology, interest in electricity and its use, financial capacity and potential of existing productivity.

In April/May 1998, Hydro Quebec and TEPCO carried out the environmental evaluation of the selected sites (MHPPs, HS and Kualeu). The main results were the following. All the sites were biophysically appropriate. However, in Bokin further geological investigations were suggested because of recent landslides. In Tendan Dua, it was recommended that the left channel be left untouched in order to protect the fish population and to avoid damage to coffee shrubs, which are a main source of revenue for the inhabitants. In Taba, an existing irrigation canal will be used as an intake canal. The water use therefore has to be carefully discussed in advance with the rice growers to prevent any conflicts. Waikelosawa has been identified as a tourist attraction by the local government. Hence special attention should be given to the architecture of the equipment in order to limit visual impact.

General recommendations include carefully taking into consideration the multiple use of water, particularly for irrigation purposes. The MHP implementation may affect rice production. As this is a dominant economic activity, it should be discussed with the farmers concerned and be appropriately compensated for. Furthermore, civil works should take place during the dry season because of reduced erosion risk and the availability of local manpower dedicated to agricultural activity in the rainy season. Beyond that, the involvement of the local population in the construction process would have economic benefits and would certainly increase their acceptance of and identification with the project. The implementation of user groups and village utilities in each village was also recommended.

The socioeconomic evaluation consists of three main activities and integrates the results of the Kualeu experience (E7, 1998a, p.29):

1. Assessment of present situation: Evaluation of the organizational potential of local communities; socioeconomic assessment study.
2. Implementation activities: Providing villagers with project information; establishing village utilities (PLD); establishing groups for interested and potential users; pro-moting income-generating activities; officially transferring all activities and assets to the community.
3. Monitoring and follow-up: Development of monitoring criteria; development of an action plan to correct deficient situations such as malfunctioning, lack of comprehension and so on.

The first activities of the socioeconomic evaluation have been implemented taking into account the Statistical Household Income Survey (SHIS) and the selection of the sites. Activity 1 is in progress and the establishment of user groups and village utilities (PLD) is planned to be finalized by March 1999. The transfer to the communities is intended to be completed in December 1999 and activity 3 might then start (ibid., p.31).

Financing

Table 5.2 Budget changes from the first to the second progress report (US$m)

	1st PR	Change %	2nd PR	E7
1. Local infrastructure/project office	0.6	– 4	0.575	}1.800
2. Micro hydro power plants (MHP)	0.7	– 16	0.585	
3. Hybrid systems (HS)	0.5	– 20	0.400	
4. Solar home systems (SHS) by E7	0.0	+ 100	0.052	
5. Other costs	0.0	+ 100	0.188	
6. Services in kind*	1.6	+/– 0	1.600	1.600
7. SHS by Indonesian government	1.0	– 90	0.100	
Total	4.4	– 23	3.5	3.4

Note: *Costs of providing consulting/engineering services including travel expenses of E7 members to RESS project.

Sources: E7 (1997, p.14) and E7 (1998a, p.7).

Table 5.2 presents the budget changes from the first to the second progress report (PR). The figures illustrate the shift of funds. On the one hand, new categories such as the SHS expenditure of the E7 and the category 'other costs' were established. On the other hand, the contribution of the Indonesian government decreased by over 90 per cent and most of the other expenditures decreased slightly too. The total budget changed from US$ 4.4 million to 3.5 million owing to the decrease in SHS, whereby the total E7 contribution remained the same. The cash distribution of US$ 1.8 million was cofunded by seven E7 members (EDF, 1997).

Time Schedule

The preparation of the RESS project started in 1993 and in December 1996 the letter of intent (LOI) and a memorandum of understanding (MOU) were signed. In Table 5.3 a comparison of the time schedules of the first and the second Progress Reports is given. It illustrates the delays which occurred during the implementation of different project components. The current time plan is to finalize the project after a two-year follow-up period which will start as soon as the technical components are implemented. Based on current information, this will be by 2001 at the earliest (Adam, 1997a, p.36).

Table 5.3 Comparison of time schedules between first and second progress reports

	1st Progress report June 1997	2nd Progress report October 1998
Solar home systems (SHS)		
Transport/installation	Start October 1997	Start October 1998 (recent information mid-February 1999)
Micro hydro power plant (MHPP)		
Waikelosawa refurbishment		
Installation, test run	Aug.–Oct. 1997	Aug. 1998–Oct. 1999
Commercial operation	October 1997	
Construction of other 3 MHPP		
Installation, test run	June–July 1998	October 1999
Commercial operation	July 1998	
Hybrid system (HS)		
Installation, test run	No date	May–August 1999
Socioeconomic integration (SEI)		
SHS		
Statistical Household Income Survey (SHIS)		
Training session	June 1997	October 1997
SHIS final report	October 1997	November 1997
MHPP		
Waikelosawa	Possibly June 1997	April–May 1998
Other MHPP		October 1998
HS	Possibly June 1997	April–May 1998

Source: E7 (1997, p.17) and E7 (1998, p.3).

PROJECT EVALUATION

The following project evaluation will be an ex ante study, mainly based on information gathered from the resident project manager C. Dauselt, the project leader H.-G. Adam (RWE) and Michael Schön from the Fraunhofer Institute of Systems and Innovation Research (FhG-ISI, Karlsruhe, Germany). M. Schön obtained his information during local investigations as an independent reviewer of the project's AIJ aspects. The external evaluation was financed by the German federal government; the main contractor was the International Solar Energy Society (ISES, Freiburg, Germany).

The evaluation of secondary objectives will mainly focus on the SHS component as there exist some ex post surveys on PV programmes in Indonesia that can be taken as an analogy. The main source has been an analysis on the impacts on the SHS in the village of Sukatani, which has been translated from Bahasa Indonesia into German (Susmarkanto and Hardianto, 1993). Furthermore, only impacts on the host country will be taken into account. This means that employment effects in the rest of the world, for instance, will not be included as this would make the analysis too complex.

Primary Objectives

Ecological objective

The ecological objective of JI is to reduce CO_2 emissions in order to combat climate change. To measure the ecological effect, the baseline scenario, that is the business-as-usual scenario, must be calculated. The baseline scenario indicates the estimated emissions without the activity. In the following, different baseline scenarios have been calculated for each of the technical components.

Solar home systems The main effect of the SHS will be the replacement of kerosene for lighting purposes in rural households. It can further be assumed that a small amount of diesel-generated and battery-stored electricity will be replaced too.

At present, lighting in rural households is provided by candles and two different kinds of kerosene lamps: so-called 'wick lanterns', simple tin cans containing kerosene with a wick inserted through a narrow tube, and 'Petromax-lamps', the usual industrially manufactured kerosene lamps.

On the basis of the results of the SHIS, it seems to be most likely that there will not be complete replacement of traditional lighting and battery use for two reasons (EDF, 1997, p.28). First, traditional lighting can easily be moved from one room to another and from outside the house to inside. If needed, the number of lanterns or candles can be increased easily and cost-effectively. Second, the various kinds of cassette-players and radios use different voltages ranging from 3V to 12V with intermediate values corresponding to the number of 1.5V batteries set in series. The old appliances will not be replaced immediately simply in order to use the new 12V electricity.

Therefore, calculations will be based on the assumption that 20 per cent of traditional lighting will continue to be operated and the low battery consumption for radio and tape-players will remain the same. These estimations are in accordance with a study undertaken in Nepal, where kerosene consumption was reduced by 79–82 per cent following the implementation of SHS in rural households (Schweizer, 1996, p.182).

Two scenarios will be calculated for kerosene consumption based on different viewpoints (see Table 5.3). In the first scenario, the kerosene consumption for lighting purposes is derived from statistics. These indicate a present average of 126.4/year consumption per household for the 'other island' region in Indonesia including NTT and South Sulawesi (GTZ and BPPT, 1995, p.27).

The second scenario is based on the type of lamp and the daily operation time. According to the calculations presented by the E7 based on World Bank estimations for Sri Lanka and Indonesia, the following assumptions are made (Cabraal and Tager, 1994, p.6): (a) one Petromax lamp with an average consumption of 0.06 litres per hour will be operated five hours a day and be replaced by a 15W bulb; (b) one wick-lantern with an average consumption of 0.04 litres per hour will be operated six hours a day and be replaced by a 6W bulb.

Both scenarios take into account that the surplus electricity which is not used for lighting purposes could be used for the operation of a small black and white TV or radio and substitute for the diesel-based battery charging (see I & II).

The yearly CO_2 offset of osn SHS according to the calculations in Table 5.4 will be in the range of 265kg (scenario I) to 407kg (scenario II), adding the diesel replacement. In contrast, the E7 have estimated an annual CO_2 emissions reduction of 522kg per SHS. This 115kg per year above scenario II. According to present E7 plans, the overall CO_2 offset of the 100 SHS, assuming a panel lifetime of 10 years, will be 522 tons. Assuming the same lifetime, total offsets under the scenarios in Table 5.4 will be 265t CO_2 or 407t CO_2 respectively. Adding the 18 SHS which were reconstructed in Kualen with a remaining lifetime of 6 years, there will be an additional CO_2 offset in the range of 29–44kg CO_2. Thus the total CO_2 emissions reduction of all 118 SHS will be about 294–451t CO_2. In the following, the conservative scenario I will be used.

For the SHS, the traditional energy consumption is taken as the reference case for the whole period of ten years. As it was one major selection criterion for the villages, it is most likely that the two selected villages will not be connected to the PLN grid in the near future. Furthermore, the evaluation of the BPPT supports this assumption. Therefore it is most unlikely that the households selected for the installation of SHS will be connected to the PLN grid in the next 10 years. If this were the case, a modification of the baseline as described for the three established MHPPs and the HS would be necessary.

Since the monetary income of the households in the villages was found to be very low (see SHIS) additional electricity consumption is not expected. Even if there were an increase in living standards, this would most likely change the consumption pattern of fuels rather than the amount of kerosene consumption. Therefore it seems unlikely that kerosene consumption for lighting purposes will increase in the near future and the steady kerosene consumption assumed above can be regarded as realistic. In order to have a more definite baseline verification over time, it is proposed to sample check a reference group (similar villages or households) during the whole period of the project.

Table 5.4 Emission reduction scenarios of an SHS

	Scenario I	Scenario II	I & II	E7 calculations	
Carbon content	'Kerosene' [a] 2.517kg CO_2/l	'Kerosene' [a] 2.517kg CO_2/l	'Diesel' [b] 1.1kg CO_2/kWh	'Kerosene' [c] 2.545kg CO_2/l	'Diesel' [c] 0.77kg CO_2/kWh
Yearly kerosene consumption/per household	126.4 l – 20% = 101 l	197.1 l[e] – 20% = 158 l	(–)	197.1 l[e]	(–)
Yearly replacement of diesel-generated electricity	(–)	(–)	9.125kWh [f]	(–)	26.918kWh [g]
Calculation	101 l x 2.517kgCO_2/l	158 l x 2.517kgCO_2/l	9.125kWh x 1.1kgCO_2/kWh	197,1 l x 2.545kgCO_2/l	26.918kWh x 0.77kgCO_2/kWh
Yearly CO_2 emissions reduction per SHS	255kgCO_2	397kgCO_2	10kgCO_2	502kgCO_2	21kgCO_2

Notes:

[a] Kerosene carbon content: 43.1 MJ/kg x 0.8kg/l x 0.073kgCO_2/MJ = 2.517kgCO_2/l (Kleemann 1994, p.XXXIV, p.XXXIII, p.233).

[b] Average overall efficiency of diesel power station in NTT is assumed to be 31.74 per cent corresponding to a specific CO_2 emission of 0.831kg/kW$_{el}$. Taking into account all other charging losses, distribution line losses and lubrication the diesel emissions will be approx. 1.1kgCO_2/kWh.

[c] E7 1997, p.30.

[d] Yearly kerosene consumption for lighting purposes: GTZ and BPPT (1995, p.27).

[e] Yearly kerosene consumption: Petromax 109.5 l/wick-lantern 87.6 l: Total 197.1 l/a.

[f] Assumption: Petromax and wick-lantern will be replaced by a 15W and a 6W bulb, operating 5 hours and 6 hours daily respectively. The daily energy production of the SHS is 170Wh. Taking a charging loss of 20 per cent of the SHS into account there remains a surplus power of: [(170Wh/d x 0.8) – (75+36)Wh/d] x 365 = 9.125kWh/a which will substitute diesel generated energy.

[g] As f but E7 did not calculate charging losses of SHS, therefore power surplus is 59Wh/d. Diesel generator efficiency is 32 per cent, charging efficiency 80 per cent diesel output is 73.75Wh/d and 73.75Wh x 365 = 26.918kWh/a (E7, 1997, p.30).

Source: Own calculations.

Micro hydro power plant A grid has to be installed to connect the households in the villages in South Sulawesi with the MHPPs. Since households used traditional energy before, the baseline will be the replacement of kerosene and battery charging as in the case of SHS (265 CO_2kg/year). Above all, this baseline will most likely take into account that the E7 has based the size of the MHPPs on the water flow instead of on the electricity demand of households in the region (E7, 1998b, p.2). Therefore it can be assumed that a considerable surplus of electricity will be produced by the MHPPs during the daytime, when households do not use electricity. In the long run, the E7 plans to connect workshops to the MHPPs to consume the electricity produced during the daytime.

All villages will have connected to the grid by 2004 according to the Second Long-Term Strategic Development Plan of the Indonesian government. Therefore a baseline change in 2004 will be assumed, taking diesel generators as the reference case up to 2020. In addition, it is assumed that by 2004 the entire electricity produced by MHPPs will be used, since the E7 will have connected consumers for the daytime electricity surplus.

At present, the electricity on the island of Sumba is generated exclusively by diesel power stations. Detailed information is available for the Waikabubak station (near the Waikelosawa site). It can be calculated, from the yearly electricity production and from the total consumption of fuel oil, that the power station has an average overall efficiency of 31.74 per cent. This corresponds to specific CO_2 emissions of 0.831kg/kWh$_{el}$. In contrast to the careful E7 calculations which assume a serviceable time of only 4380 h/year the calculations here are based on 6000 h/year. On the basis of inspections by M. Schön and information from C. Dauselt, this serviceable time seems to be a realistic figure for a hydroelectric run-of-river station utilizing a quite steady stream flow like the one at the Waikelosawa site. Since more detailed information about the other sites is missing, the same baselines and serviceable times are assumed from 2004 onwards. The corresponding annual CO_2 offsets of the different sites are listed in Table 5.5.

Table 5.5 Annual CO_2 offsets of the MHP sites

	Taba	Tendan-Dua	Bokin	Waikelosawa	Total
Plant capacity	46kW	64kW	13kW	15kW	138kW
Annual energy production[a]	276 MWh	384 MWh	78 MWh	90 MWh	828 MWh
CO_2 offset					
2000	47t CO_2	50t CO_2	14t CO_2	75t CO_2	186t CO_2
2001	75t CO_2	80t CO_2	22t CO_2	75t CO_2	252t CO_2
2002	93t CO_2	100t CO_2	28t CO_2	75t CO_2	296t CO_2
2003	93t CO_2	100t CO_2	28t CO_2	75t CO_2	296t CO_2
2004–20[b]	229t CO_2	319t CO_2	65t CO_2	75t CO_2	688t CO_2
E7 annual CO_2 offset [c]	155t CO_2	216t CO_2	44t CO_2	51t CO_2	466t CO_2

Notes:
[a] Calculation: Plant capacity x annual serviceable time of 6000 h/year.
[b] Annual energy production x diesel CO_2 emission rate of 0.831kg/kWh$_{el}$.
[c] Capacity x serviceable time (4380 h/year) x diesel CO_2 emission rate (0.77kg/kWh$_{el}$) (E7, 1998a, p.43).

Source: Own calculations.

The resulting overall CO_2 emissions reductions are approximately 12Mt CO_2 for the four systems over an expected lifetime of 20 years or 19Mt CO_2 over a lifetime of 30 years. In contrast, the E7 estimated an 8 per cent lower CO_2 offset using only diesel generators as the reference case as well as a much lower annual serviceable time and a lower diesel CO_2 emission rate.

Hybrid System In this study we assume the same reference case for the HS as for the MHPPs in South Sulawesi described above. In contrast the E7 calculations are again based on the assumption that, without the HS, only diesel-generated electricity would have been consumed.

To calculate the reduced emissions from 2000–2004, the traditional energy consumption will be assumed as the reference case, since no electricity was available before. For the remaining 16 years the electricity produced by the solar panels and the wind generator has to be estimated. Since no other information for solar radiation and wind is available at present, the calculations in Table 5.6 are mainly based on E7 figures.

For emergency operation a zero CO_2 emission reduction can be assumed. Hence the diesel generator of the project can be considered as equal to the diesel generator of the reference case. The resulting CO_2 offset of the HS is estimated to be 581t CO_2 over an expected lifetime of 20 years. The offset calculated by the E7 was 600t CO_2 (4 per cent more) for the same period.

Table 5.6 CO_2 emission reduction by the HS

	Capacity	Solar radiation/ Wind	Capacity factor	Electricity produced
PV panels	20kW	3.4h/d	1.00	24 820kWh/year[a]
Wind generator	10kW	24h/d	0.16	14 016kWh/year[b]
Total electricity				38 836kWh/year
CO_2 offset				2000 (80hh): 21.2t CO_2 2001–3 (100hh): 26.5t CO_2/year 2004–20: 30t CO_2/year [c]

Notes:
[a] Calculation: 20kW x 3.4h/d x 365d = 24 820kWh/year.
[b] Calculation: 10kW x 24h/d x 0.16 x 365d = 14 016kWh/year.
[c] Calculation: 38 836kWh/year x 0.77kg CO_2/kWh = 30t CO_2/year.

Source: E7 (1998a, p.44).

Finally, the overall CO_2 offset of the project is calculated and presented in Table 5.7. The overall project will lead to CO_2 emissions reductions in the range of 12 884 to 12 913t CO_2. Considering the lack of information about the real lifetime of the technical components another estimation has been undertaken presuming a 30 year lifetime for MHPPs. The resulting total CO_2 offset of the MHPPs based on this lifetime is about 18 718t CO_2.

Table 5.6 Total CO_2 offset of the project

	SHS	MHP	HS	Total
Number	100 (18)	4	1	
Technical specification	a 50Wp	138kW	20kW PV and 10kW Wind	
Model timeframe	10 (6)	20 (30)	20	
CO_2 offset	265t CO_2 (29t CO_2)	12 038t CO_2 (18 718t CO_2)	581t CO_2	12 884t CO_2

The significant differences between the two CO_2 offsets calculated for the MHPPs demonstrate the importance of an ex post evaluation based on the real conditions and lifetimes of systems. The calculations presented above can only give a rough idea of the range of CO_2 emission reductions which can be expected from the different systems.

Economic objective

The economic objective of CDM/JI is to reduce CO_2 emissions more cost-effectively than at home. Therefore the costs per ton of CO_2 reduction will be estimated and compared with CO_2 reduction costs in the E7 countries. It is assumed that the E7 will pay for all technical components and services. Only fees to sustain the systems will be collected from the users to cover O&M costs. The present cash flows were calculated by the E7 and indicate that sufficient revenues will be collected from the villagers to sustain the SHS and the three MHPPs. The Waikelosawa site will begin to have positive cash flows in year six of operation, but the HS will not be economic over the whole period (E7, 1998a, p.32). To simplify the calculations, it is assumed that all components will meet the self-financing target and no further financing by the E7 is needed.

The real discount rate will be assumed to be 4 per cent based on the E7 opportunity costs. Taxes can be excluded from the calculations, as agreed by the Indonesian government (E7, 1997, p.12). Moreover, it is necessary to distinguish between direct project-related costs and transaction costs (including for example, search, negotiation, approval, insurance, monitoring and evaluation costs) and to calculate the costs for different time periods.

In view of the lack of information about transaction costs, three different options will be calculated. The first option is based only on investment costs and search and implementation costs and can therefore be considered to be the lowest level of reduction costs. Since the second option takes into account all investment and transaction costs, this will most likely indicate the highest costs. Finally, in the third option, the long term transaction costs for the SHS will be estimated, based on a Scandinavian analysis. This means that transaction costs will be calculated assuming a well-functioning CDM. Thus transaction costs will be reduced significantly in line with learning effects gained during the pilot phase. As it has not yet been decided how the CDM will be implemented and organized, this option has been divided into two parts. Part A compiles all costs that have to be paid by the investor. Part B also considers the project-related costs of the CDM authorities. It is assumed that the negotiations will take place bilaterally. Furthermore, it is assumed that crediting, verification and documentation has to be made by an independent expert from the investing country. In part B, monitoring and verification costs are not regarded as transactions costs, both for calculating convenience and because these costs are usually not directly financed by the project initiator. Table 5.8 reflects the reduction costs per ton of CO_2 for each project component and for the entire project. It has to be emphasized that the calculations are only rough estimations owing to the lack of information. More information about the assumptions can be found in the annex.

Table 5.8 CO_2 reduction costs in thousands US$ per ton of CO_2 (NPV in thousands US$)

Project component	Option I		Option II		Option III	
					A	B
SHS	1.4–1.0	(–302)	4.3–4.7	(–1 255)	0.91–1.04 (–267)	1.36–1.51 (–400)
MHP	0.06	(–735)	0.11	(–1 306)	-	-
HS	0.86	(–500)	1.52	(–881)	-	-
Total	0.119	(–1 537)	0.27	(–3 442)	-	-

Sources: Table 5.7 and the annex to the present chapter.

Table 5.8 demonstrates that reduction costs range from US$60/t CO_2 to US$4700/t CO_2, depending on which costs and which components have been included in the calculation. Furthermore, the figures show that the lowest reduction costs can always be related to the MHPPs and that SHS have the highest reduction costs for Option II. The latter has to be related to the high transaction costs in the preparation phase according to the original plan of installing 1000 SHS. Above all, Option III indicates that reduction costs for SHS will significantly decrease in the future.

Comparing these results with reduction costs in the investing countries causes difficulties. Hence, usually, no proper indications are given in studies as to the costs (transaction costs or investment costs) on which the calculations are based. Taking into account that search and negotiation costs will be much lower in the country of origin, it is most likely that there will be more reasonable abatement options for instance in Germany than in the RESS project.

Other objectives

Other objectives of AIJ are the additionality and sustainability of AIJ projects. This means that AIJ: 'should bring about real, measurable and long-term environmental benefits related to the mitigation of climate change that would not have occurred in the absence of such activities' and 'that the financing of activities implemented jointly shall be additional to the financial obligations of Parties included in Annex II to the Convention within the framework of the financial mechanism as well as to current official development assistance (ODA) flows' (FCCC/CP/1995/7/Add.1, Decision 5/CP.1).

To prove additionality an investigation of barriers which prevent the transfer of capital or technologies must take place. Unless such barriers can be identified, there is no reason to believe that emission reduction can only be related to AIJ. Considering the problems that the implementation of renewable energy technology systems faces on a worldwide level, it can be stated that the RESS project complies with the technological additionality criteria. The financial additionality is only fulfilled if the entire project is financed independently by the E7. That means it will fail if the Indonesian government provides any SHS because these SHS would have been installed anyway.

Sustainability refers to the required long-term character of projects. Whether the project was really implemented in a sustainable matter can only be proved at the end of the estimated lifetime, but will most likely be fulfilled if the criteria of low project failure risk and participation of local authorities and people are met.

Regarding the RESS project, it is most likely that the project risk, which includes technical, economic and political risks, will be relatively low. In terms of the technical risk, one can state that the SHS and MHP technologies are already approved in Indonesia and a guarantee is given by constructing firms and producers. In contrast, the hybrid system, the design of which is quite complex (including three different technologies) would have a relatively high technical risk even if robustness and local operators were among the key design guidelines. Considering the economic risk, the E7 has already accepted the negative cash flows of the investment, particularly of the HS, and increases in construction costs are usually included in the construction subcontracts. Therefore this risk can be ignored. In terms of the political risk, the resident project manager estimated it to be rather low, taking into account that the project has already been approved by the Indonesian government as an AIJ pilot project.

The criterion of local participation is fulfilled, taking into account the establishment of village utilities and user groups, as well as the training of local technicians. This training will include the definition of the different roles the users and technicians have to play, which is regarded as a key issue in a sustainable project implementation. In addition, the financial sustainability analysis (FSA) will also help to achieve this long-term target by getting rural households to pay a monthly rate to cover operating costs and to build up a reserve fund for substitute facilities. Finally, the cooperation with the active regional NGOs will improve the project's sustainability.

Secondary Objectives

In the following an analysis of the indirect and direct secondary impacts of the RESS project is presented. As previously mentioned, these estimations are ex ante projections and will mostly focus on the impacts of the SHS component, as more information is available for this technology, as well as stronger impacts being assumed. Furthermore, it was not always possible to estimate indirect effects. Above all, it must be kept in mind that the results of the SHS impact analysis for Sukatani (a relatively rich village in central Java) cannot be transmitted entirely on to the selected villages in NTT because of the significant difference in standards of living.

Ecological objectives

The ecological objectives consist of impacts on the local environment and on resources. The major local environmental side-effect of the RESS project will be the improvement of local air quality owing to the reduction of SO_2, NO_x and particulate matter. Since air pollution is not a severe problem in these regions of NTT, with their low population densities, these effects have to be considered as minor side-effects. On the other hand, the improvement of indoor air resulting from the installation of SHS will be a more important benefit, since the fumes from burning kerosene cause chronic respiratory irritations. Reducing the number of small batteries will have marginally positive effects on mercury and lead concentrations in soil and zinc concentrations in soil and water. As the solar panels, micro hydro systems and wind generators will be very small, and the density and number of installations low, it can be assumed that side-effects owing to aesthetic impacts on landscape or noise (wind generators) will be negligible.

Drawing up a reference scenario for the cumulated resource consumption of the different project components would correspond to a comparison of life-cycle assessments. This would be a very complex analysis and would go beyond the aim of this chapter. Therefore, in the following, only some of the points will be examined in detail. Regarding the land area needed it can be concluded that the area for the installation of SHS will be rather small because they will be pole-mounted. In contrast, the MHP stations will cover more land, constructing powerhouses, penstocks and headraces. As indicated in the second progress report, the HS will need a land area of 400m^2 (E7, 1998a, p.19). This has to be compared to the land area used by the reference diesel power plant.

Furthermore, battery use should be estimated. In the business-as-usual scenario, the households were using car batteries for running TVs. After the installation of the SHS it might be assumed that the lifetime of the batteries (now used for SHS) would be extended (by 50 per cent) because of the utilization of adequate charge controllers. Thus a slight positive effect on resource usage can be expected of SHS. Indirect positive effects in this area might occur from the recycling of the batteries, but in a country where no recycling system for batteries currently exists this is a rather difficult aim.

Economic objectives

As indicators of the direct and indirect economic impacts of the project, the effects on the balance of payments, employment, productivity, technology transfer and innovation in the host country will be analysed. The project will have a direct impact on the balance of payments, increasing foreign currency and capital in the host country by at least US$1.8 m. As some of the components will be imported, these costs must be subtracted. In addition, foreign currency might be saved as a result of the replacement of kerosene and diesel, which are mostly imported and also highly subsidized. Thus implementing renewable energy systems would contribute to reducing foreign debt and decreasing public expenditure.

The construction work and the management/maintenance of the different systems and the overall project will presumably create full-time and temporary jobs in the host country. The construction of the different components will have a local short-term employment effect. The establishment of PLDs (chairman, vice-chairman, cashier and spare-part manager) and the training of technicians has led to new part-time jobs. Furthermore, founding the project office in Kupang has created long-term full-time jobs. Besides the resident project manager, about five local people have been employed. One is working as a driver and the others are undertaking office work or cleaning or are responsible for security. It is difficult to quantify the net direct employment effects of the whole project. For instance, at this stage no information is available on the number of people involved in the construction of some components. Also it has to be assumed that all the employees were unemployed before working for the RESS project, otherwise they would have to be subtracted from the total number of employees. Positive indirect employment effects might be assumed in the forward and backward linked industries, for instance in enterprises producing technical components for any of the systems. In contrast, some indirect negative effects on traditional kerosene or diesel suppliers as well as on Petromax lamp producers (wholesalers and retailers in Indonesia) might be expected.

Positive impacts on productivity will result mainly from the new electricity supply in the SHS villages. The electricity might be used for the extension of working hours or for new production opportunities as well as for an increase in information through electrical appliances (TV, radio). Also production losses might be reduced as a result of independence from fuel suppliers. As experience in the village of Sukatani has shown, these effects are often overestimated. Electricity is mostly used for consumption purposes such as lighting and entertainment. The only effects on production in Sukatani were the extension of the opening hours of some shops and an increase in crab meat production. However, the latter was mainly caused by increasing demand, and not by the availability of electricity.

The number and the size of new projects using the same technology or, for instance the establishment of a PV industry owing to the increase in demand could be an indicator of technology transfer resulting from demonstration effects. As these effects cannot be linked precisely to one project, demonstration effects are very difficult to estimate, particularly ex ante. Nevertheless, it can be assumed that the project will have marginal positive effects on technology dissemination and transfer.

It may be assumed that the project will have marginal positive direct and indirect educational impacts. Training provided by the E7 will allow direct educational benefits for people involved in the administration, construction and maintenance of the systems. Indirect educational effects can be achieved by electricity improving learning conditions. Pupils might finish their homework in the evening (after Koran school) or read a book using the better lighting conditions. As has been outlined by the Sukatani study, mosques and Koran schools were extending their opening and praying times as a result of the use of solar-powered lighting. In addition, thanks to media-related electrical appliances, access to information will improve. This is especially important for women because they hardly ever leave the village. The benefits related to the access to new media sources depends to a great extent on the level of education and the quality of programmes broadcast.[3] Above all, the knowledge gained simply by operating electrical appliances such as radios and TVs must be also considered.

Finally, the impact on innovation in the host country has to be analysed. As previously mentioned, the SHS users in Sukatani have been replacing the common bulbs with 2W bulbs from cars. This demonstrates an impact on creativity and innovation. Indirect innovative effects, like impacts on project-related technologies or the increase of innovative creativity, are difficult to measure. However, as experience in Sukatani shows, some people were using solar energy for experimentation. A 15-metre radio station and an electrical welder have been created. Taking these impacts into account, marginal positive direct and indirect effects on innovation might be the result of the RESS project.

Sociocultural objectives

In order to estimate the secondary sociocultural effects of the RESS project, capacity-building factors, impacts on awareness creation, distribution effects and effects on the standard of living will be analysed. The number of established institutions and infrastructure (telephone, fax, Internet) will be used as an indicator to quantify the direct impacts on capacity building. In connection with the RESS project, the establishment of the project office in Kupang and of PLDs in the villages and the creation of networks with the Indonesian government and local NGOs will be counted. These institutions and networks will help to reduce the transaction costs of future projects such as CDM projects. Regarding the communication infrastructure, only marginal effects, such as the purchased telephone, fax and Internet facilities of the project office, can be seen so far.

In order to estimate influences on the cognitive and affective environmental awareness of the local population, it would be necessary to question the SHS

3. The resident project manager stated that the TV programme (TVRI), which a typical rural household in NTT can receive, broadcasts mainly political propaganda and fantasy movies. Therefore information value and educational effects are assumed to be low. However, this may have changed as a result of recent political changes and of the fact that there was already a tendency to more critical reports, for example on health issues, from 1995.

users ex ante and ex post of the system installation. Their habitual way of living in their natural environment has to be investigated. Such an examination would certainly go beyond the aim of this chapter. We shall therefore refer to the estimations of an expert, who quantified these effects as marginal (Betz, 1997, pp.120, 170).

Direct distribution effects will be analysed according to different population groups. First, the rural population will benefit from the SHS, and the gap between rural and urban areas may be diminished. Second, direct income effects will arise owing to the above-mentioned employment effects and the increase in productivity. Third, distribution effects within the households may occur. Thus women and children tend to profit the most from the SHS. Women may do their housework in the evening and use the daytime for other things. Children have the possibility of doing their homework in the evening and will be the main users of TV and radio (Cabraal and Tager, 1994, p.7). Positive indirect income effects will arise in companies that are involved in the production or sale of project-related components. Negative impacts are assumed for the traditional suppliers of kerosene and diesel. Furthermore, social conflicts may be the result of the installation of SHS because of the prestige increase of the selected households. To avoid social conflicts over unfair use of electricity and to preserve social harmony, which is also important for the sustainability of the project, electricity consumption meters have been installed in each household connected to MHPPs and HS and the tariffs will be adjusted respectively (E7, 1998b, p.4).

Above all, major impacts on living standards and habits are expected to result from the project, possibly overlapping with other effects. Direct effects will include impacts on health, leisure and other habits, as well as income effects. The quantification of income effects has to include, besides the above-mentioned employment related effects, savings owing to the lower maintenance costs of SHS compared with traditional energy use. However, if the E7 appropriate these savings as a monthly fee no direct monetary benefits will be left. In contrast, time savings can most likely be gained owing to the fact that there is no longer any need to take batteries to recharging stations (usually two trips for delivery and pick-up). To quantify net time saving, one has to subtract the time for the maintenance of SHS and measure it according to the personal income level. Furthermore, the increase in comfort has to be taken into account as well as the independence from traditional energy suppliers. Another main benefit has to be considered concerning health conditions, namely the improvement of indoor air quality, which reduces the risk of chronic respiratory irritation. Additional improvements are the reduced fire hazard in the wooden cottages and fewer burn injuries, as well as benefits to the eyes from better lighting. Finally, electricity will increase the living standard in the villages and improve the entertainment options. As an indirect positive impact, this can help to reduce migration from rural to urban areas. The impacts on the standard of living can be illustrated by the experiences gained in Sukatani. Before the introduction of SHS in 1989, the population owned 12 TV sets and 22 radios or cassette-players. Four years later, in 1993, the number of TV sets had increased to 61, the number of radios and cassette-players to 80, and seven Karaoke facilities were available. As the standard of living in the villages in NTT is much lower, these impacts will not be as strong as in Sukatani.

Final Evaluation

The Evaluation Matrix (Table 5.9) summarizes and evaluates the primary and secondary objectives and impacts of the project. It demonstrates that the secondary impacts will most likely surpass the primary impacts to a great extent. Conflicts between primary and secondary objectives are not very likely to occur.[4] Negative results have been estimated for the cost of emissions reduction. The same amount of money invested in abatement options in other countries and other technologies could have achieved higher emissions reductions. As calculated above, the emissions reductions of the overall project (12 884t CO_2) will be relatively small compared to the high costs of the project (US$3.4 m.). High reduction costs are to be found, especially with the hybrid system, which can be ranked as the most inefficient both in terms of emissions reduction per kWh and in costs of electricity generation. It is most likely that the HS is only a showcase and would never have been constructed under real conditions in this specific conception, taking the technological risk and the negative cash flows into account. The SHS rank first in terms of emissions reduction per kWh. Therefore the decrease to 900 from the original 1000 SHS had a significant effect on the overall emissions reduction of the project. Regarding the costs of electricity generation, the SHS rank after the MHPPs.

To sum up, according to the technologies examined, micro hydro seems to be the most economic renewable energy generation option for rural areas in Indonesia. But as long as micro hydro plants mainly replace electricity produced by diesel generators they will not have such a significant CO_2 emissions reduction impact as the SHS. Therefore, it would be important to establish all systems in regions where grid connection seems least probable in

4. A big hydro power station AIJ project (with a dam) might lead to conflicts between primary and secondary objectives because it could for example, destroy the flora and fauna and cause other damage.

the long term. As there are extremely high reduction costs, it can be assumed that the same project would not have been implemented under an established CDM system. Further, it has to be ensured that no SHS financed by the Indonesian government will be included in the project, as this would not fulfil the criterion of avoiding financial additionality (see p.112).

Table 5.9 Evaluation matrix

Primary objectives	Impact	Indicator	Estimated effect
Ecological	Greenhouse gas emissions reductions	Reduced CO_2 emissions through replacement of traditional kerosene or diesel consumption	+
Economic	Emissions reduction costs	Costs compared to reduced emissions	-
Others	Additionality	Investigation of barriers	+
		Independent financing of the project	?
	Sustainability	Project risk	+ (relatively low)
		Participation	+
Secondary objectives			
Ecological	Local environment	Other emissions concentrations	+
		Aesthetic impact on landscape	+ (relatively low)
		Noise production	+
	Resources	Soil use	+/–
		Consumption of resources (battery lifetime)	+

Economic	Balance of payments	Net foreign currency and capital inflow (including replacement of imported kerosene and diesel)	+
	Employment	Direct: net creation of long-term/short-term and full-time/part-time jobs	+
		Indirect: creation of jobs in forward and backward linked industries	+
	Productivity	Productivity increase in SHS, MHP, HS villages	+
	Technology transfer	Dissemination of PV, hydro or wind technology	+
	Education	Direct: number of trained people	+
		Indirect: general educational options (homework, reading books, radio, TV)	+
	Innovation	Direct: modification and adaptation of technology components	+
		Indirect: increased creativity owing to electricity supply	+

Sociocultural	Capacity building/ infrastructure	Number and size of established institutions and infrastructure (communicational and other)	+
	Awareness	Behaviour change because of increased knowledge	?
	Distribution	Direct: profit distribution over different groups of society, decrease in rural /urban disparities	+
		Indirect: income effects on forward and backward linked industries, social conflicts	+
	Standard of living	Direct: health conditions, leisure time, entertainment options	+
		Indirect: decrease in rural migration	+

In the following, some problems concerning the accuracy of the emissions reduction calculations will be analysed.

The two different scenario calculations for SHS (see p.104) indicating emissions reductions ranging from 265kg CO_2/year (scenario I) to 407kg CO_2/year (scenario II) demonstrate the sensitivity of the results. First, accurate estimations of the average carbon content of Indonesian kerosene and diesel are necessary. Second, an analysis of traditional energy consumption, for instance under the SHIS, would have made it much easier to calculate the appropriate emissions reductions. To solve the latter problem – which will be even worse when evaluating projects with longer lifetimes – the monitoring of the development and energy consumption of a reference group (reference villages or households) as a sample check could be a reasonable solution, as stated above.

Furthermore, the problem of life-cycle emissions should be taken into account to ensure that the project leads to net GHG reductions and not to an increase in net emissions. Regarding the SHS, a calculation of the emissions caused by the production and transport of the SHS components (solar modules, batteries and so on) has to be made and compared with the emissions from the refining and transport of kerosene. Such calculations will increase the costs of baseline verification and should therefore only be roughly estimated.

The problem of so-called 'transaction emissions', that is emissions for instance from project-related travel must also be considered. As an example, the following estimation was made. If an expert travels from Germany to Indonesia (NTT) the CO_2 emissions can be assumed to be 2.7t CO_2. This corresponds to the annual CO_2 reductions of between seven and ten SHS.[5] According to the RESS project, E7 experts have travelled more than once from Europe to Indonesia. Thus these emissions should not be underestimated. As previously stated, for economic reasons these emissions should only be estimated and compared with overall emissions reductions to get an idea of the volume and the net reductions.

Moreover, leakage effects should be considered. In connection with the SHS, a positive side-effect could be the demonstration effects which can lead to a reduction of the barriers (such as a lack of information) that have so far prevented the growth of solar technology use.

Reducing transaction costs, which were extremely high in the RESS project even taking into account all 'first-mover problems', would lead to a significant total cost reduction. This might have been possible if there had been cooperation with experienced institutions right from the beginning. In the RESS project, cooperation with development aid institutions such as the GTZ (German Agency for Technical Cooperation), which already has good connections with the Indonesian government and established networks, would have been more than beneficial to the E7.

Regarding Indonesia's development strategies which were analysed in the introductory sections of this chapter, the RESS project seems to be compatible with the socioeconomic and environment priorities of the country and will support its development plans. In this context the criticism is justfied that Indonesian politics is highly inconsistent considering the subsidies to kerosene, diesel and electricity which distort the incentives for energy use and renewable energy production. However, comparing Indonesian policy with other national policies, it definitely has to be taken into account that

---

5. A total distance of 27 000km including the international return flight and domestic flights to the sites was assumed. Further, the calculations are based on a kerosene consumption of 0.0318kg/km per person and a specific emission factor of 3.15kg CO_2/km. (All information received in personal communication from a pilot to M. Schön.)

there have been no import taxes on SHS components as is the case in other developing countries (Cabraal and Tager, 1994, p.18). Finally, it would be better to select countries for CDM projects which have only marginal energy price distorting policies or none at all.

Several factors have to be mentioned which have been responsible for the failure to meet the time schedule during project implementation. The Asian financial crisis and the political changes in Indonesia are two of the causes. However, the delays had already appeared before these problems arose. On the one hand, complex structures and changing responsibility in the Indonesian government led to delays in approving the project as an AIJ project. On the other hand, the E7 experts involved could not give the RESS project the high priority it would have needed according to the time schedule. Moreover, the fact that major changes in the project design have to be approved by the steering committee of the E7 makes rapid decision making difficult. Therefore more transparent structures, both in the host country and on the part of the investor would be a marked improvement.

Transparency is also important to ensure that projects are only listed on one AIJ or CDM country list and that future double counting such as that which happened in the RESS project can be avoided.[6] After the pilot phase, problems can occur in deciding which country or investor may claim the credits for the emissions reductions. Therefore these problems have to be sorted out in advance. In the RESS project, the credit sharing has not yet been sorted out.

In spring 1999, the RESS project was only listed on the German AIJ project list, not on the UNFCCC list, as official approval by the German and Indonesian environmental ministries had not yet been granted. According to direct information from the person responsible in the E7, this formal act will be finished by mid-1999, after the elections in Indonesia.

In addition, the reporting on projects should be transparent and easily accessible to ensure that the public is well informed. Thus 'window dressing' and 'image making' with projects where no results can be seen so far should be avoided (see RESS project[7]). However, misunderstanding between media and project initiators may lead to this kind of divergence in reporting. In this context it should be mentioned that, currently the only incentive to carry out any AIJ projects is to gain first-mover advantages and to use the projects for publicity purposes. As long as there is no crediting for emissions reductions,

---

6. The RESS project was listed on the German AIJ project list (BMU, 1996, p.1) and was mentioned at the same time in a Japanese press release as a Japanese AIJ project (TEPCO, 1996).
7. In one newspaper, some photos of the implemented SHS of the RESS project were shown, whereas in reality the sites for the SHS had not even been selected (Merkel, 1996, p.352).

these will remain the only incentives. Therefore it has to be acknowledged that, under these conditions, at least the E7 has invested in an AIJ project. This helps more experience to be gained in the pilot phase, as only a very few project ideas have been implemented so far. Furthermore, it can be assumed that this will support the wider distribution of renewable technologies.

Finally, it has to be noted that the E7 was setting itself high targets, especially as regards socioeconomic integration. This particularly has to be recognized since the E7 is still a young organization, only founded in 1992. The several surveys undertaken, as well as the established project office, demonstrate its great efforts, as does the choice of the residential project manager, whose knowledge and experience gained from his former activities for the World Bank and the GTZ in Indonesia in rural energy projects was very important.

SUMMARY

The evaluation of the RESS project demonstrated the large number of objectives that have to be achieved within a CDM project to ensure that it fulfils both the climate change (CDM, primary objectives) and development related criteria. If the latter are not met, this means the project does not have secondary benefits for the developing country and is not compatible with the country's development strategies, in which case the project will most likely not have been approved by the host country. Therefore the best solution would be to have project categories which include only projects fulfilling most of the primary and secondary criteria and indicating rough estimations of baselines. This would, on the one hand, ensure a high acceptance by both the host country and the investors and, on the other hand, be beneficial in reducing transaction costs. Currently, one can only speculate as to what kind of projects will be listed, but it seems most likely that renewable energy technologies will be included. Regarding the reduction costs, which will even decrease in the future, the RESS project, for instance, has demonstrated that micro hydro power seems to have a potential for AIJ projects in rural areas in Indonesia.

The experience with the RESS project has shown that there is still a considerable lack of knowledge regarding AIJ or JI and CDM. In particular, knowledge of procedures and the relevant information needed to carry out such a project and to fulfil the given criteria are missing. I therefore propose putting a source of information (booklet or information bureau) at the interested company's disposal which will serve as a guide from the beginning to the end of the process.

Finally, it can be concluded that it is necessary to solve the problems which have occurred within the AIJ pilot phase before starting the CDM. Therefore an overall evaluation of the pilot phase would be useful. The definitions of modalities and procedures for the CDM should consider the following problems and solutions that have been discussed in more detail in the present chapter: reference groups, life-cycle emissions, transaction emissions, leakage effects, transaction costs, failure to meet the time schedule, double counting and proper reporting. If all these problems have been examined, and a decision on how to handle them is taken which will apply uniformly to all projects under the CDM, this flexible instrument may be introduced and will be beneficial for combating both climate change and underdevelopment.

REFERENCES

Adam, Hans-Georg (1997a), 'Joint Implementation II. Die E7-Initiative', EUmagazin, (4), 36.

Adam, Hans-Georg (1997b), 'Project Updating Form', Essen.

Adam, Hans-Georg (1999), 'Experience of RWE AG and the E7 Initiative with AIJ Projects', workshop on 'Project Types for Flexible Instruments-the Situation after Buenos Aires', January, Karlsruhe.

Betz, Regina (1997), 'Joint Implementation: Ein Instrument im Dienste von Klima- und Entwicklungspolitik?' Eine Studie am Beispiel des 'Regenerativen Energiesystem Projekts' der E7-Initiative in Indonesien, Fraunhofer ISI-Workingpaper, Karlsruhe.

BPPT (1995), Draft Project Aid Proposal for the Fifty Megawatt Peak Photovoltaic Rural Electrification Programme in Indonesia. 1994 to 2004, Jakarta.

Bundesministerium für Umwelt, Naturschutz und Reaktorsicherheit (BMU) (1996), Activities Implemented Jointly. Deutsche Pilotprojekte, August 1996, Bonn.

Bundesstelle für Außenhandelsinformationen (1994), *Indonesien, Energiewirtschaft*, Cologne.

Cabraal, Anil and Carolyn Tager (1994), 'Solar Photovoltaics: Best Practices for Household Electrification', draft for comment, Asia Alternative Energy Unit (ASTAE), World Bank, Jakarta.

E7 (1997), *Renewable Energy Supply Systems in Indonesia*, Progress Report No.1, Essen.

E7 (1998a), *Renewable Energy Supply Systems in Indonesia*, Progress Report No. 2, Essen.

E7 (1998b), *Indonesia – Renewable Energy Systems* (AIJ), Summary Notes of Project Team Meeting, December 1998, Essen.

EDF (1997), *Report on Statistical Household Income Survey*, Paris.

GTZ (Deutsche Gesellschaft für Technische Zusammenarbeit) and BPPT (Agency for the Assessment and Application of Technology) (1995), *Technology Assessment for Energy Related CO_2 Reduction Strategies for Indonesia*, Eschborn.

Kleemann, Manfred (1994), *Energy Use and Air Pollution in Indonesia*, Aldershot.

Merkel, Angela (1996), 'Ein außerordentlich interessantes Klimavorsoge-Instrument', Energiewirtschafltiche Tagesfragen, No. 6/vol. 46, pp.352–3.

OECD (1996), *Main Economic Indicators*, July.

Schweizer, Petra (1996), *Psychologische Faktoren bei der Nutzung regenerativer Energien: Eine Studie zum Einsatz von Solartechnik im Zentralen Himalaya,* Lengrich.

Susmarkanto, Drs. and Ir Amy Hardianto, (1993), *Analisis Dampak Sosial, Ekonomi dan Budaya. Dalma Rangka Pengkajian dan Penerapan Pembangkit Listrik Tenga Surya di Desa Sukatani*, Jakarta.

TEPCO (1996), 'AIJ Project to Reduce Greenhouse Gas Emissions', http://www.tepco.ca.jp/corp-com/press/96071101-e.html, 9.7.1997.

ANNEX 5A1

Figure 5A1.1 *Indonesia*

ANNEX 5A2

Table 5A2.1 Option I in thousands US$

Period	Type of costs	SHS	MHPPs	HS	Total
Preparation t_{-1}	PC/TC[a]	240	144	96	480
	Search	120	72	48	240
	Feasibility	120	72	48	240
Implementation t_0	PC[b]	52	585	400	1 037
	Investment	52	585	400	1 037

Notes:
[a] PC = project costs; TC = transaction costs
[b] Investment costs (see E7, 1998a, p.7).

Table 5A2.2 Option II in thousands US$

Period	Type of costs	SHS	MHPPs	HS	Total
Preparation t_{-1}	PC/TC[a]	240	144	96	480
	Search	120	72	48	240
	Feasibility	120	72	48	240
	TC	309	186	123	618
	Negotiation	206	124	82	412
	Approval	103	62	41	179
	Insurance	–	–	–	–
Implementation t_0	TC	360	216	144	720
	Implementation	360	216	144	720
	PC[b]	52	585	400	1,037
	Investment	52	585	400	1.037
t_1	TC	144	86	58	288
	Monitoring/Evaluation	144	86	58	288
t_2	TC	144	86	58	288
	Monitoring/Evaluation	144	86	58	288

Notes:

[a] PC= project costs; TC= transaction costs.
The E7 project leader (Hans-Georg Adam) in September 1997 estimated the total transaction costs to be US$2.2 m., with the following distribution over the components: 50% SHS, 30% MHP and 20% HS. Based on estimations by the author there will be a marginal increase in these costs owing to the extended project implementation period. Therefore calculations are based on the following assumptions:
Total transaction costs: US$2.4 m.
SHS (50%): US$1.2 m. ($t_{-1}$: 0.46; t_0: 0.3; t_1: 0.12; t_2: 0.12)
MHP (30%): US$0.7 m. ($t_{-1}$: 0.46; t_0: 0.3; t_1: 0.12; t_2: 0.12)
HS (20%): US$0.5 m. ($t_{-1}$: 0.46; t_0: 0.3; t_1: 0.12; t_2: 0.12)

[b] Investment costs (see E7, 1998a, p.7).

Calculation of option I:

$$\text{Net Present Value} = \frac{(-PC/TC)t_{-1}}{(1.04)^{-1}} + (-PC)t_0$$

Calculation of option II:

$$\text{NPV}=\frac{(-PC/TC-TC)_{-1}}{(!.04)^{-1}} + (-PC-TC)t_0 + \frac{(-TC)t_{-1}}{(1.04)^{1}} + \frac{(-TC)t_{-2}}{(1.04)^{2}}$$

Table 5A2.3 Option III (only SHS)

Period	Costs	Bilateral project	Option III A & B (US$)	Financed by
All periods	Administration	7% of PC	3 640	CDM authority
Preparation t_{-1}	PC/TC[b]		2 080	
	Search/ feasibility [a]	4% of PC [b]	2 080	Investor
	TC		65 604	
	Negotiation[c]	0.25 PM[d]	4 260	Investor
	Approval[e]	3.00 PM	51 120	Investor
	Crediting[f]	0.5 PM+ TE	10 224	CDM authority
	Insurance	–	–	Investor
Implementation t_0	TC		155 000	
	Implementation		155 000	Investor
	PC[g]		52 000	
	Investment		52 000	Investor
	Operation & Maintenance		–	Investor
t_1 to t_{n-1}	TC		10 520	
	Monitoring[h]	6 000 US$ [c]	2 000	CDM authority
	Documentation /Crediting[i]	0.5 PM	8 520	CDM authority

Notes:

[a] The search costs, which are the costs of finding interested partners, will be added to the practicability costs. The overall costs in the preparation period will be approximately 4 per cent of the investment costs in a foreign host country and only 2 per cent within the investing country.

[b] PC= IC = project costs, here the same as investment costs because there are no operational costs; TC= transaction costs.

[c] Negotiation costs involve the costs of coming to an agreement between host country and investor.

[d] Research personnel is calculated at approx. 100 000 DKK, which corresponds to US$17 040 per person month (PM). 1DKK = US$0.1704 on 28.6.1996 (OECD, Main Economic Indicators, July 1996, p.199).

[e] Approval costs are the costs when negotiated exchange must be approved by a government

agency. They consist of an application, which may involve a detailed description of the project and a rough estimation of the baseline; 3 PM will be assumed for the generation of this information.

[f] Crediting will be carried out under the guidelines of the CDM authority. Half a PM and travel expenses (TE) of US$1 704 will be assumed.

[g] The investment costs of the SHS are equal to the contract sum of US$52 000.

[h] Spot checks will be undertaken every three years to monitor and verify the emissions reduction and the estimated baseline. The costs of a spot check are assumed to be about US$6000 causing total fix costs of US$2000 per year.

[i] It is assumed that, for yearly documentation and crediting, 0.5 PM has to be spent.

Sources: Own calculations and Nordic Council of Ministers Joint Implementation of Commitments to Mitigate Climate Change – Analysis of 5 Selected Energy Projects in Eastern Europe, TemaNord 1996:573, Stockholm/Oslo. (1996, p.58).

Calculations:

Option III A: $NPV = \frac{-57460}{(1.04)^{-1}} + (-207000)$

Option III B with an expected lifetime of n=10.

$$NPV = \frac{(-67684-3640)}{(1.04)^{-1}} + (-155000-52000-3640) + \frac{(-10520-3640)\cdot[1-\frac{1}{(1.04)^{10}}]}{0.04}$$

6. Host Country-driven Implementation: the Case of Costa Rica

Michael Dutschke[1]

SUMMARY

This chapter discusses the first results of the AIJ pilot phase in Costa Rica. Costa Rica has a relatively high level of economic and social development and a well-developed environmental policy which is comparable to that of advanced industrial countries. It is a major destination for ecotourism. Nevertheless, it suffers from severe deforestation owing to the unequal distribution of land, to migration and cattle ranching as well as the expansion of coffee and banana plantations. Moreover, transport emissions are rising rapidly and fossil fuel electricity generation is growing fast, despite the country's target of phasing out fossil fuels completely.

Costa Rica's knowledge base is high and capacity building almost unnecessary. Thus Costa Rica was able to develop creative environment policy instruments such as debt-for-nature swaps and biodiversity prospecting to attract foreign funding. It is not surprising that it was the first developing country to open a JI office, develop project approval criteria and host AIJ pilot projects. The conditions for project-based climate cooperation in Costa Rica are excellent compared with the average developing country.

Nevertheless, the success can at best described as mixed. Only a third of the projects are actually funded, though several of them seem to be profitable even without a value for carbon. Most of them are proposed by US entities. To attract more funding, the JI office now certifies tradable mitigation bonds and encourages multi-sector large-scale projects where transaction costs are lower and coherence with national development objectives can be more easily checked. It directs its attention to public investors such as the USA and Norwegian governments. The renewable energy projects suffer from the

1 Hamburg Institute for Economic Research (HWWA), Neuer Jungfernstieg 21, 20347 Germany, Phone +49 403562479, Fax +49 402991855, E-mail: dutschke@hwwa.uni-hamburg.de.

unrealistic target of phasing out fossil fuels by 2001, thus making AIJ projects in this sector impossible from that time onward. The bulk of the projects therefore concern forestry which is prone to uncertainties in the calculation of emission sequestration. A comparison of the estimates shows wildly differing assumptions in the baselines and sequestration capacity of the forests. Whether actual project implementation conforms to the plans remains to be seen. An independent verification of project results is currently being undertaken.

The analysis of the Costa Rican case shows that CDM can only be successful in the long run if the industrial countries offer incentives to investors and if baseline determination rests on a clear set of guidelines. Human and technical capacities are necessary but not sufficient conditions for successful climate cooperation projects in developing countries. They seem to be able to prevent complete project failures, though, and can lead to innovative approaches. The issue will only be settled if large-scale project investment is forthcoming. The ability to process huge numbers of project proposals and check whether they conform to development priorities, as well as monitoring and verification, will then become crucial.

COSTA RICA AS AIJ HOST COUNTRY

Costa Rica is perhaps the most active host country in the current pilot phase of Activities Implemented Jointly (AIJ). As it was the first to break the blockade of the developing countries organized in the Group of 77, it received much criticism, but for the same reason it had the 'first-mover' advantage among the developing countries. It received eight out of the 15 first round US projects.

In order to evaluate the Costa Rican AIJ policy, a brief introduction to the country's geographic conditions and ecological problems is presented below. Special attention will be paid to tourism, which made ecology an export factor for Costa Rica.

Environmental Conditions and Policy

Costa Rica forms part of the geographical bridge both between North and South America and between the Atlantic and Pacific oceans. It is among the ten countries with the highest amount of precipitation, but the precipitation patterns differ from region to region. A remarkable variety of climatic regions can be found within its territory, thanks to the mountain range that ascends from sea level to nearly 4000 metres.

This geographic complexity is reflected in Costa Rica's huge biodiversity. Although Costa Rica represents only 0.035 per cent of the earth's territory, scientists estimate that between 3 and 7 per cent of all species live within its boundaries (for example, Fuchs, 1997, p.38). On the other hand, the extreme weather conditions lead to erosion problems. Over 60 per cent of its territory is unsuitable for agriculture (LeBlanc, 1997, p.2).

Compared with other countries of the subcontinent, there is a remarkable public awareness of environmental matters in Costa Rica. There are a number of reasons for this. Although the social climate has become tougher in recent years, poverty is not as big a problem as it is in other parts of Latin America. Since the army was abolished in 1949, a major share of public budgets has been accorded to social welfare. Since 1949, no coup has taken place. A peculiarity of the Costa Rican democratic system is the *consensualismo*, which means 'embracing the political opponent'. Thus non-governmental organizations (NGOs) find themselves integrated in the fulfilment of state functions, namely in nature conservation and environmental politics. The right to a healthy environment was laid down as a constitutional amendment in 1994 (Saborio Valverde, 1997, pp.8, 38).

Protected areas now cover nearly 25 per cent of national territory. Eleven per cent of the territory belongs to the strongest protection category, which is the declaration as a national park in the property of the state. Once declared, the status of a national park cannot be withdrawn, not even by law. Problems arise between the protection of nature and the constitutional protection of private property (Art. 45). In some cases where compensation was only paid in state endowments the High Court granted full property rights to private landowners within the boundaries of national parks (SGS, 1998, p.28). Specialists distinguish seven legal statutes, depending on the type of owner (private, NGO – willing or unwilling to transfer property – state organization) and the progress of litigation. Only 5.4 per cent of the lands in national parks and biological reserves are registered as state-owned (ibid., p.14; Goldberg *et al.*, 1998, p.5). As a consequence, the owner can be tempted to make the most of the land while it still belongs to him or her and fell the remaining valuable trees in a short space of time. This is why the Costa Rican government has desperately been seeking funds for buying ground within these 'paper parks'.[2]

The environment ministry, MINAE,[3] was founded in 1990. Ecology was made a compulsory subject at primary school. Classes undertake excursions to the national parks to learn about nature conservation. Sustainable development was a major policy goal of the government of President José

2. The question is whether there is no legal alternative to this enormous commitment of state capital.
3. Ministerio de Ambiente y Energía, created by law 7152 on 21 July 1990. Its current name was bestowed in November 1995.

María Figueres. The Costa Rican Planning Ministry created a powerful system of sustainable development indicators (SIDES)[4] which allowed the fulfilment of this civil right to be measured. It includes social, economic, ecological and climate data. Figueres, in spite of being the son of the founder of the republic and in spite of his recognition abroad for his environmental commitment, was quite unpopular in his own country. People doubted his ability to really improve their living conditions. The newly elected government of Miguel Rodriguez undertook a process of *concertación* which includes environmental services (*servicios ambientales*). A parliamentary commission and a public forum on the topic tried to define the ecological benefits of forestry subsidies and how these should be financed (CSA, 1998; FNC, 1998). Recently, environmental groups have been complaining that the current government seems unwilling to follow the proposals made by the *concertación* (Escofet, 1999).

Deforestation and forestry policy

At the time it was discovered by the Spaniards, 95 per cent of Costa Rican territory was covered by different kinds of forests. Although logging had already been begun by the indigenous pre-Columbian people and then proceeded under the Spanish colonists, it was not until the second half of the 20th century that massive lumbering destroyed the largest part of the virgin forests. Projections in the year 1992 foresaw complete deforestation between the years 2015 and 2033 (Notimex, 1993, p.21). The latest figures published by the *Ministerio de Planificación* indicate a decrease in deforestation from 17 000 hectares in 1992 to 8000 in 1994 (MIDEPLAN, 1997). Nowadays about 1.8 million hectares of primary forest remain, most of which is under some kind of protection (for 1994, see MIDEPLAN, 1997).

The reasons for deforestation, apart from the tropical timber business, are (a) the expansion of coffee and banana plantations, (b) export-oriented beef production and (c) the displacement of subsistence farming to less productive areas not claimed by the big land-owning companies. The beef export boom which started in the 1960s offered opportunities even for smaller farmers because investment needs are low owing to the fact that cattle are kept outside all year. Most of the beef is produced for export. Because of the high prestige of cattle farming – the 'cattle subculture' (LeBlanc, 1997, p.18) – high incentives are required to counteract its expansion. Each kilogram of beef produced implies the loss of 2.5 tons[5] of soil (Fuchs, 1997, p.29). This results from the fact that half of the national territory (Lara, *et al.*, 1995, p.116; Santiago and Schmidt, 1994, p.2) is covered by pastures, while only 8 per cent of it is regarded as suitable for cattle grazing (Lara, 1995, p.116). Much

4. It can be accessed on the internet (URL: http://www.mideplan.go.cr/sides).
5. Metric units are used throughout this chapter.

of the farmland lies on steep hillsides and in areas where tropical rainfall easily washes away the soil. On the other hand, beef exports only account for 1.4 per cent of export revenues (for 1996, see MIDEPLAN, 1997), while the World Resources Institute estimates the value of natural resources lost between 1970 and 1989 to be US$4.1 billion (Tenenbaum, 1996, p.17).

Subsistence farming on marginal land is a result of the unequal land distribution. Costa Rican law tries to compensate for this at the expense of primary forests: after two years of occupying land, the *'precaristas'* (squatters) are given possession rights; after ten years, the 'squatters' can claim a property title (LeBlanc, 1997, p.10). The vicious circle consists in the fact that previously tropical woodlands are rapidly exhausted and degraded by erosion, as roots no longer hold the soil together. Farmers thus see themselves forced to clear more virgin forests.

Forestry activities have been subsidized since 1979. For reforestation measures, a partial exemption from capital tax was available. Its benefits went only to the big landowners to whom this tax was applicable. From 1986, the *Certificados de Abonos Forestales* (CAF) give the right to exemption from any tax in the first five years of reforestation up to the amount of the total costs. This subsidy is equivalent to about US$1000 per hectare. But tax exemption still excludes small landowners who usually do not pay any direct taxes.

In 1988 a revolving forestry fund, the *Fondo de Desarrollo Forestal* (FDF), was created in the context of a debt-for-nature agreement with the Netherlands. It was meant to encourage small forestry, crediting US$644 per hectare for the first five years, which were to be repaid as the wood was harvested. The FDF has now been dissolved (Heindrichs, 1997, p.5). The CAF certificate system was remodelled in 1991 and linked to the pursuit of a sustainable forest management plan for each section of forest. Access was made easier for small landowners (introduction of the prepaid CAFa – *CAF por adelanto*). A tax reduction on wood-selling benefits from 30 per cent to 20 per cent is granted if the area has been reforested. This is not an instrument to prevent the logging of primary forest, nor is the incentive for reforestation strong enough to take effect (LeBlanc, 1997, p.3).

These instruments did create incentives for reforestation, but they showed some significant shortcomings.

1. Primary forests were logged in order to allow tree planting, because most subsidies only referred to reforestation and planting.
2. The certificates were traded by investment companies that did not care about the long-term protection of the new forests.
3. Forest management restricted the number of species to be planted. Most indigenous trees were excluded because there were no data available on

growth and output (Butterfield, 1994, p.319). Of the subsidized lands 60 per cent were reforested with exotic trees (Heindrichs, 1997, p.7).

Between 1979 and 1995, forest plantation measures on 167 451ha were subsidized to the amount of US$101.2 million (ibid., p.6). Only in 1992 were two subspecies of CAF *(CAFma por el manejo del bosque* and *CAFma 2000/CPB Certificado por la Protección del Bosque)* introduced for the protection of existing forests (ibid., p.5).

The 1996 forestry law instituted the Forestry Environmental Services Payment Programme (FESP) in a stepwise replacement of the CAF.[6] The FESP includes the Private Forestry PFP for subsidies to private landowners and the Protected Areas Project PAP aimed at purchasing land within the 'paper parks' (Subak, 1998, pp.9f).[7] However, a new certificate was introduced for forest conservation (*Certificado de Conservación de Bosque* – CCB) (Heindrichs, 1997, p.8). Responsibility was transferred to the now restructured *Fondo Nacional de Financiamiento Forestal* – FONAFIFO (LeBlanc, 1997, p.3). The municipalities became entitled to logging allowances (Muñoz, 1996, p.2), surveyed by private forestry engineers. Under certain conditions, they are even entitled to allow the logging of 'intervening' primary forests once in a while (*bosque intervenido*). The regionalization led to a 16 per cent increase in logging permits (Quesada, 1997, p.6A). The area covered by permits more than doubled, from 431 566 to over one million cubic metres (Escofet, 1997, p.12). This is only partly owing to the fact that permits are now being granted for several years in advance. The main problems are the lack of employees and skills in the regional administration and the increased temptation to accept bribery. Critics state that Costa Rica lacks a systematic measuring of its forest cover (ibid.) The estimation that '50 per cent of the logging in Costa Rica is done without the required permit' (LeBlanc, 1997, p.3) can neither be proven nor denied. Logging cannot even be impeded within national parks because the administration – the *Sistema Nacional de Áreas de Conservación* (SINAC) – does not have enough rangers to control the areas. Yet there is great public awareness of the issue. Lately, the abolishment of lumber shipment tags was retracted because of the large number of complaints about transports of allegedly illegally cut-down trees (Escofet, 1997, p.12). The new Costa Rican government now plans to tie logging allowances to certification by non-governmental organizations (Escofet, 1998). Although Costa Rican forestry policy has been trying to attribute a material and social value to forest cultivation and protection, its

6. New CAF applications were not accepted from this point, with the exception of small landowners (with under 10 ha), for whom the CAF was prolonged for 10 more years (Heindrichs, 1997, p.8).
7. The concept of environmental services will be dealt with later on in this chapter.

means are far too complex for the potential beneficiaries. It lacks constancy and reliability. There is still too much emphasis on planting compared with the protection of the existing forests. The Costa Rican idea of allowing the sustainable management of primary forests, even if it does not lead to serious environmental damage, contributes to confusion in defining the status of the last remaining virgin forests.

Agriculture and environment

Plantations, which were first established in the middle of the 17th century, have become predominant in agriculture. The first crop was cocoa, followed later by coffee in the high plains and bananas in the Caribbean lowlands. Organic waste is one problem. Some 70 per cent of the total organic products of coffee and 40 per cent of banana plants are dumped, in many cases into the rivers, a practice which has been illegal since 1938 (Oakes, 1996). Coffee and banana growing are often linked to the abusive use of pesticides, which currently lies seven times above world per capita average (Saito and Odenyo, 1997, p.2) thereby endangering farmworkers' health and the water resources. The massive use of agrochemicals in the cultivation of banana and new non-traditional crops (tropical fruit, macadamia nuts and flowers) and tannic acids stemming from coffee processing are major threats to ground and surface water. Since a new sun-resistant coffee bush was first planted in 1980 the trees which provided shade have been cut down, leading to the disappearance of 90 per cent of the birds living in coffee plantations and to higher soil erosion (Oakes, 1996). Recently, much attention has been paid to reducing hazardous use of pesticides in banana plantations. After years of lawsuits, Dow Chemicals offered an out-of-court settlement to the banana workers affected by sterility after using DBCP (Avalos Rodríguez, 1997, p.8A), a pesticide banned in the USA since 1979 (Saito and Odenyo, 1997, pp.2ff). Chiquita Corp. and other banana producers have tried to install an own ecolabel which is certified by the New York Rainforest Alliance and the Costa Rican *Fundación Ambio* (Anonymous, 1997b; Anonymous, 1997c). Critics object that the 'better bananas' principles more or less simply reflect the requirements of the law on solid and liquid wastes enacted in January 1995 (Scharlowski, 1996).

Population and settlement structure

Population growth endangers sustainability as well. Although the density of 67 inhabitants per square kilometre does not indicate overpopulation, the habitable part of the territory is relatively small, and two-thirds of the 3.4 million population lives in the central valley. Population grows by annual rates of around 2.5 per cent, 3.2 per cent in urban areas. This causes problems owing to inadequate infrastructure, air pollution and settlements competing with the agricultural use of the country's most fertile soils.

Energy production and policy

Today's share of renewable energy sources is 82.4 per cent, 75.2 per cent alone stemming from the Arenal hydroelectric plant. The first block of the geothermal plant, Miravalles, has been operational since March 1994; the second one was intended to go on-line in the middle of 1997 (ICE, 1997; Cordero, 1996), but it has not been finished yet. An operator for the third block has been found by tender, using the BOT (build–operate–transfer) model for a span of 15 years (Cordero, 1996; Segnini, 1997). In 1994 the minister for natural resources, energy and mining (MINAE) promised a phase-out of fossil energy production by the year 2001. The officials of the national JI office (OCIC) are very unhappy about this prematurely set aim because it is by no means rational. First, it assumes that electricity consumption remains constant. In fact, power demand is growing by approximately 8 per cent per year (OCIC, 1997, p.1). Second, there is not yet a technically viable substitute for burning fossil fuel or gas in the quantities needed during peak load. Finally, El Niño also led to a shortage of water supplies in Costa Rica.

The Energy Savings Act of 1996 introduced a 15 per cent tax on all fossil fuels. It provided for one third of the revenues to go to the national forestry fund FONAFIFO. As the Ministry of Finance considered itself unable to fulfil the requirement of contributing US$15 million to the fund in 1996, in 1997 both institutions agreed to guarantee FONAFIFO an annual 2.7 billion colones[8] for five years. The centre-right Rodriguez government elected in March 1998 has already unveiled plans to abolish the fuel tax (Escofet, 1999).

Industry and environment

Industry is relatively backward concerning waste management and energy efficiency. Although 47 per cent of industrial energy demand is already met by the use of agricultural waste, 37 per cent still stems from burning fossil fuel. Case studies for five typical Costa Rican enterprises by the German Agency for Technical Cooperation (GTZ) found large potentials for the reduction and replacement of energy use. Put into practice, the proposed changes could lead to an annual climate benefit of the order of 4000 tons of CO_2 and 80 tons of SO_2. At the same time, cost saving potential would range between 10 per cent and 13 per cent (GTZ, 1996, p.25).

---

8. This amount is equivalent to US$10 million at the present time. According to the agreement, it is secured against falling below US$7 million as a result of devaluation (personal communication by Paulo Manso, OCIC, 11 July, 1997).

The wood processing industry still receives its raw material at low prices from settlers, which does very little to motivate them towards sustainable forestry. On the other hand, cheap raw material leads to squandering. The nature conservation NGO *Fundación Neotrópica* estimates that only 54 per cent of the logged wood reaches the sawmill, where half of the wood delivered is wasted by unproductive processing (Butterfield, 1994, p.318).

As far as CO_2 emissions are concerned, the highest growth occurred in the transport sector: 'From 1983 to 1993, the number of vehicles in use doubled from 190,000 to 390,000, with the number of automobiles increasing from 66,000 to 150,000' (LeBlanc, 1997, p.4): Recently, the Costa Rican government eliminated the 40 per cent consumption tax on electric vehicles in order to make them more competitive. The enforced use of electric vehicles in public and private transport could in the long run ease pollution in urban areas and at the same time lower the CO_2 emissions caused by transport (Muñoz, 1997, p.3), given the structure of electric power production described above. However, if electric vehicles do reach important numbers, this will lead to even higher increases in electricity demand.

Tourism

In its boom years at the end of the 1980s and the beginning of the 1990s, tourism grew by 25 per cent to 30 per cent annually. It was in 1994 that foreign currency earnings generated by tourism surpassed those from banana exports for the first time. Today tourism employs 17 per cent of the active population (Burkard, 1996, p.20). Most tourists come from the USA and Latin America (41 per cent each). The share of ecotourists to Costa Rica is estimated to be above 40 per cent (Panos Institute, 1996). The term ecotourism or sustainable tourism is not clearly defined. At any rate, ecologically aware travellers choose the country because of its natural beauty and would not visit it if it did not have them. Among these travellers, scientific visitors or birdwatchers can be found, as well as whitewater rafters who tend to consider nature a setting for their recreational activities. Worldwide ecotourism has the fastest growing market share within tourism with a growth rate of between 10 per cent and 15 per cent (ibid.). After an initial phase of scepticism, the government of Costa Rica decided to support ecotourism strongly. Then environment minister Carlos Rösch advocated the integration of a clause concerning sustainable tourism into Agenda 21. The average ecotourist spends more money on his or her vacation than normal tourists. In 1995 every traveller to Costa Rica left US$840 within the country. This figure rose by 71 per cent within only eight years, which reflects a rise in quality. Ecotourism relies very much on regulation in order to prevent itself destroying its own bases. The tourism ministry gives priority to small and medium-sized enterprises and takes care that their projects are benign to the environment. In 1993, for instance, a German investor was expelled because of irregularities in the construction of a hotel complex. He was charged with having eradicated valuable vegetation, killed animals belonging to protected species and damaged a coral reef (Anonymous, 1993).

The village of Longo Mai is an example an ecologically and socially oriented tourism development. It lies next to La Amistad National Park in the south-east mountain region near the Panamanian border. Longo Mai has specialized in hosting educational travellers and solidarity workers (Burkard, 1996, p.21). The regional effects of tourism can also be studied in La Fortuna, near the Arenal volcano. The village resembles a gold digger town, not only by its name. Practically all of its 800 inhabitants are in one way or another involved in tourism. The typical hostel does not exceed five rooms or cabins. Small supermarkets and restaurants provide food for the visitors; handicraft gift shops offer guided tours to the volcano and the hot springs beneath it.

The national park entrance fees for foreigners were raised overnight from US$1.3 (Panos Institute, 1996, p.10) to US$8, which at the same time both increased the tourists' participation in conservation and halted the crowding of the areas. In some places now the number of visitors at any one time is limited. Small farmers and land workers find jobs as rangers or guides in the parks, thus ensuring the support of the local population for nature conservation. Tourism also has its share in Central American cooperation. Guatemala, Belize, El Salvador, Honduras, Nicaragua, Costa Rica and Panama are planning the installation of a 1500 mile biodiversity corridor, the *Paseo Pantera* (Stevens, 1996).

Climate Cooperation in National Politics

The external financing of nature conservation policy is not new to Costa Rica. It started with the so-called 'debt-for-nature programmes', which cancelled external debts under the condition that their value, converted into national currency, was to be invested in national parks. Another creative means of generating income from the preservation of primary forests is their use as a resource for genetic material. The first time their exploitation was put on a regular basis was in 1991, with the cooperation between the USA-based pharmaceutical enterprise Merck & Co and the National Biodiversity Institute INBio, founded as an NGO by the Costa Rican government in 1989. The agreement involves the systematic collection and documentation of samples from the rain forests by INBio specialists and their use by Merck's research laboratories. Merck paid about US$1 million as a fixed sum on every renewal of the agreement.[9] A licence fee of between 1 and 3 per cent (in some cases

9. The initial payment is said to have been US$1.14 million. There have been two renewals up to now, one in 1993, the other in 1997.

up to 10 per cent) of the revenue is granted if drugs based on Costa Rican genetic substances go on to the market. The gene database is made available to the public by the company via the Internet (URL: http://www.merck.com). INBio is to use only half of this income for its operative purposes while the rest goes to the preservation of the forests (Tenenbaum, 1996, p.19). The rights of the author stay within the country (Anonymous, 1997b). This contract, the full text of which is kept secret, has been widely criticized for selling out a country's natural resources. Still it has to be considered that the 'traditional' way to generate income is by destroying the resources. Costa Rica attempts to preserve its natural beauty as a way to have its cake and eat it too: whole sectors of the economy are based on the survival of the indigenous flora and fauna.

On the basis of its prior experience, Costa Rica was among the first countries to play an active role in AIJ, taking the chance to promote its sustainability policy. It offers all the necessary requirements for successful climate cooperation: strong and lasting democratic institutions and wide acceptance of the goal of climate protection. Joint measures may help to counteract the pressures of exploitative industry and farming. The measures financed do not induce a new path of development, they just foster and stabilize a process already begun by Costa Rican politics. Climate cooperation can set further economic incentives towards realizing the value of nature's resources. In contrast to the theory of joint implementation, this is not a transfer of know-how from north to south. In the field of nature conservation Costa Rica is establishing knowledge which is transferable to other tropical regions.

The concept of environmental services

Environmental services has become the key term for ecologically motivated subsidies. The 1996 forestry law differentiates between four service categories: carbon fixation, watershed protection, biodiversity and ecosystem protection, and protection of scenic beauty. Recent legislation added the protection of soils (CSA, 1998). The financial mechanism behind environmental services is the Forestry Fund, FONAFIFO. It collects and allocates transfers from parties who benefit from nature to those who participate in the protection and restoration of nature. These are mainly the national park system, forest owners who conserve or replant forests and the Biodiversity Institute, INBio. There is a growing group of payers into this fund: foreign park visitors who pay higher entrance fees, pharmaceutical companies who investigate genetic resources, and fuel buyers. Thus foreign investment in greenhouse gas (GHG) mitigation is not the only source of income for the FONAFIFO. Costa Rican governments have always been very cautious about questioning the ecological role of the big US banana companies. Only in March 1997 did a MINAE study reveal the damage banana plantations do to the environment (Alvarado Davila, 1997). This is a step beyond the traditional understanding of sovereignty in Central America. In a contract disclosed in September 1998, the Ministry for the Environment MINAE and the transnational fruit company DelOro agreed to the transferral of 1200 ha of land to the state-owned Guanacaste Conservation Area (GCA) as a payment for five kinds of environmental services received by DelOro (Dutschke, 1998, pp.25f):

1. 'Biological control agents' (insects) coming from the GCA to the plantations.
2. Technical services rendered by national and international GCA consultants to DelOro
3. Water provision from the nearby Rio Mena.
4. Biodegradation of 1000 annual truckloads of orange peel.
5. Protection of an isolated orchard which serves to produce new seedlings in case pests destroy the other plantations.

Each of these services is mentioned with price per unit and a minimum fee. Pricing seems somewhat arbitrary though, being bound to match exactly the value of the lands to be transferred. Any use of these services beyond the quantities mentioned is subject to further payment. On the other hand, carbon benefits resulting from restoring forests on the transferred lands will be shared between DelOro and the GCA. This clause speculates on the valuation of these benefits on a future international mitigation market.

The Costa Rican JI office and its guidelines

Costa Rican AIJ cooperation started in September 1994 with a 'Statement of intent for bilateral sustainable development cooperation and joint implementation of measures to reduce emissions of greenhouse gases' (USIJI, 1994), followed a year later by a similar document on cooperation between the USA and all Central American states. The aims of cooperation are best described in the bilateral statement. The following items are explicitly listed, but are not intended to be exclusive: 'biodiversity conservation and ecosystem protection, reduction of local pollution, sustainable land-use practices, improved rural income opportunities, and local participation in project planning and execution' (ibid.).

In April 1994, the Costa Rican JI Office (*Oficina Costaricense de Implementación Conjunta*, OCIC) was created as a result of the cooperation between the later Ministry for Environment and Energy (MINAE),[10] the

---

10. In 1994, its name was *Ministerio de Recursos Naturales, Energía y Minas* (MINIREM).

privately organized Costa Rican Investment and Trade Development Board (CINDE) and two NGOs. One is the FUNDECOR, an NGO dedicated to nature conservation, whose president, Franz Tattenbach, at the same time heads OCIC. The other is the ACOPE, the Association of Independent Power Producers. OCIC receives additional funding from CRUSA, the Costa Rican–US Foundation for cultural exchange. OCIC was established by a presidential decree. It reports to the MINAE and executes the authority to formulate AIJ policy, and evaluates and approves projects (LeBlanc, 1997, p.7). Nonetheless, it is not located clearly within the system of separation of powers. OCIC consists of only seven persons, most of whom constantly represent the institution in international meetings. Five of them are scientists.

The OCIC's guidelines for project criteria are as follows: minimize red tape, be experience-based, meet current international standards, represent Costa Rica's particular interests and address GHG abatement benefits sharing among participants (UNFCCC, 1998b). Project proposals are to be decided upon within six weeks. There are different sets of criteria, which can be grouped into general criteria, climate priorities and feasibility items.

The general criteria state that projects should be in accordance with Costa Rican laws and sustainability goals. They should offer 'enhancement of income opportunities and quality of life for rural peoples and members of certain vulnerable groups including cultural minorities' (Lay *et al.*, 1996). The communities involved have to support the project. The transfer of skills and technology is desired as well as that the negative influences of the project be kept to an acceptable level.

Criteria cited from the UN Framework Convention on Climate Change (FCCC) are reinforced by OCIC and exceeded insofar as verification by a 'qualified, non-participating organization' (ibid.) is requested. Financial additionality to development assistance or any other obligations by industrialized countries is called for, according to the UNFCCC. All costs related to the project have to be considered, including those of non-participants. Institutional feasibility is demanded on the Costa Rican side. Political, administrative or scientific institutions must be able to administer the project as well as the proponent. Previous experience in climate cooperation on the proponent's part are highly appreciated.

CURRENT AIJ PROJECTS IN COSTA RICA

Nine bilateral AIJ projects have been approved in Costa Rica. The total amount to be invested in forestry projects is ten times higher than in energy projects, and its volume in terms of carbon offsets is expected to be 73 times higher than in the energy projects (Gorbitz, 1997, p.55).

Of the initial five forestry projects approved by USIJI, only one project is fully financed. Another one received just enough funding to realize a pre-feasibility study (and was subsequently removed from the list). Two out of four energy plants are definitely operational. The lack of finance is as typical for the AIJ pilot phase as the fact that all participants tend to hide this fact from the public. A blatant example of this behaviour is the latest report to the FCCC, in which projects that are not financed are described as being operational. Cooperation with Norway is special because projects are not only approved but also financed by the guest country's JI body.

The project descriptions below are structured as follows: once the objective is outlined, participants are listed, climate effects and project costs are calculated and possible externalities are taken into consideration. At the end of each description, additional information and a short summary will be given.

Forestry Projects

Regarding climate effects, there are three possible forestry project forms. One consists of the preservation of existing forests in order to prevent adding GHG from deforestation to the emissions from the combustion of fossil fuels. This means that there is no imminent emission reduction. The second is the restoration of degraded forests. Degradation takes place when the density of species in one area is not high enough to sustain the existing forest cover. This depends on various factors, such as precipitation levels, erosion and the vegetation zone (Lund, 1998). The third alternative is tree plantation on land where vegetation cover was previously lower. This last distinction is important, because plantation policy risks being a zero-sum game if existing forests are cleared for the purpose. In addition to that, the resulting biodiversity losses can be irreversible.

In theory, constant reforestation and plantation could for a time sequester emissions from industry and traffic, thus winning time for a change of patterns in production and use of energy. The problem is that pests, fire or simply logging and changes in land use can reverse the progress made in all the years of forest growing. In contrast, emission reductions, once achieved from efficiency gains, cannot be reversed in the future. The aspect of future losses makes forestry projects difficult to handle in the context of climate cooperation. Formerly, two projects, BIODIVERSIFIX and CARFIX, had been proposed. Because they lacked finance, they were removed from the FCCC listing and included in the Protected Areas Project described below.

ECOLAND/Esquinas National Park

ECOLAND is the acronym for Esquinas Carbon Offset Land Conservation Initiative. The 13.4 square km Esquinas delta is located on the Golfo Dulce, opposite the Osa peninsula in the extreme south-west of the country. In 1994, it was declared a national park under the name of Piedras Blancas. The project aims to purchase nearly 20 per cent of the privately owned park area and to transfer it to the national park administration. All except 350 hectares of the total of 2500 hectares to be bought are currently forested (UNFCCC, 1997). Although the commercial participants will not be linked to the operation of the park, ECOLAND is the only conservation project to be fully financed. It started in January 1995 and has a life-span of 15 years.

Participants ECOLAND is managed by Trexler and Associates, Inc. (TAA), a USA-based consultancy, which specializes in AIJ operations. Astonishingly, since the 1997 report to the FCCC, TAA has no longer been mentioned. In the ECOLAND case, MINAE cooperates with five US enterprises and three NGOs. Tenaska Washington Partners, Ltd is the managing partner among four different power companies. The NGO participants include the US National Fish and Wildlife Foundation, the private Costa Rican forest conservation foundation COMBOS (*Conservación y Manejo de Bosques Tropicales*) and *Regenwald der Österreicher*, an Austrian non-profit organization which supports an ecotourism project bordering on the park (USIJI, 1996, pp.63f) and will also provide for monitoring the project (UNFCCC, 1997). According to USIJI, ECOLAND was among the first projects initiated by the Costa Rican government in December 1994.

Climate effects The baseline calculation assumes, that under normal circumstances, the area would be completely deforested within 15 years. 'Some landowners hold logging concessions, a number of which are active, and many owners face economic pressures that encourage deforestation' (USIJI, 1996, p.64). There is no further explanation of this statement. The calculation refers to 'general soil and vegetation carbon content literature' (ibid., p.65). The numbers cited in this context are put in italics (see Table 6.1). The fixation loss prevented is added to fixation gains from the project, which makes a total carbon offset of 366 200 tons.

Different fixation quotas, varying according to decade and vegetation zone, are not considered, which can be explained by the relatively small project area. ECOLAND is the only project to take carbon fixation in the soils into account. Very little is known about the annual growth of the so-called 'humus soils' in the tropical rainforest. Numbers between 0.8 and 5.1 tons per hectare have been found (Nilsson and Schopfhauser, 1995, pp.267ff). In one Costa Rican case after 18 years of regeneration, no significant growth of humus soil could be proven (Herold, 1995, p.30). Refraining from speculation on humus growth would result in lowering the estimates by 44 per cent.

Of the total carbon offsets, only 250 000 tons are 'credited' to Tenaska. This is done at once, although annual results are no higher than 23 037 tons. This does not matter during the pilot phase, but will create a liability problem as soon as GHG crediting becomes effective within the CDM system.

Table 6.1 Total carbon sequestration in ECOLAND

Reference scenario	
Total extension	2 500ha
Without forest	350ha
Forested area	2 150ha
Deforestation time	15 years
Annual deforestation	135.73ha
Soil fixation/ha	*143t C/ha*
Released by logging	*60%*, equivalent to 75t C/ha
Carbon fixed in vegetation/ha	*110t C/ha*
Released by logging	*80%*, equivalent to 88t C/ha
Complete Carbon fixation	235t C/ha
Released by logging	163t C/ha
Annual loss	23 363t C
Total loss in 15 years	350 450t C
Project scenario	
Annual growth fixation/ha	*3t C/ha*
Growth extension	350ha
Annual fixation	1 050t C
Total gain in 15 years	15 750t

Sources: USIJI (1996, pp.65ff.), UNFCCC (1997), own calculations.

Costs Land purchases cost US$910 000, which is about US$380 per hectare. An endowment of US$40 000 for annual implementation costs was created. Project development and representation, that is transaction costs, were US$150 000 (UNFCCC, 1997) or 14 per cent of total project costs, which were covered by Tenaska. The project turns out to be quite cheap, with the net costs per credited ton of carbon coming to US$4.4, and even cheaper for Tenaska when discounting the US$450 000 contributed by the NGOs.

Externalities Contrary to the project approval criteria of both national JI bodies, no information is given about the side-effects. The following information is missing:

- How many owners are there within the newly declared national park?
- Is the land inhabited?
- Which part of the park area has already been logged?
- What made developers choose the current project area?

Observations It is remarkable that only the ECOLAND developers came up with the idea of calculating carbon fixation in soils, thereby nearly doubling the expected climate effect. Social and regional economic effects have not been taken into account. It would be interesting to know where the money goes. Will it generate work or will it migrate to the capital? Another problem is the lack of independent third-party monitoring.

KLINKIFIX reforestation project

KLINKIFIX was approved in November 1995. It stands for carbon fixation as fast and as efficiently as possible. The fast-growing pine species Klinki (araukaria hunsteinii) is planned to be cultivated on former marginal pastures in forestry plantations. The conifer originates from Papua New Guinea and produces wood suitable for utility poles or plywood (WBCSD, 1997). It is to provide a new source of income for the farmers who market their carbon benefits in a kind of joint venture with the main project developer. The projected lifetime has lately been extended from 40 to 46 years, six of which count as the implementation phase (see Table 6.2). Implicitly the carbon calculations indicate that between years 41 and 46 nearly all the trees will be cut down. Project location is the Turrialba Valley, a 30km beeline east of the capital. Although there is no substantial funding yet, the project is reported to have started in June 1997 (UNFCCC, 1997).

Table 6.2 Klinki project case

Year	Annual growth (hectares)	Cumulated (hectares)	Annual growth (t C)	Cumulated (t C)
1	100	100		
2	500	600	820	820
3	1 000	1 600	4 920	5 740
4	1 300	2 900	13 120	18 860
5	1 525	4 425	23 780	42 640
6	1 575	6 000	36 285	78 925
7	0	6 000	49 200	128 125
41	0	6 000	49 200	1 800 925
42	–649	5 351	47 944	1 848 869
43	–991	4 360	43 881	1 892 750
44	–1 288	3 072	35 755	1 928 505
45	–1 511	1 561	25 191	1 953 696
46	?	?	12 799	1 966 495

Source: UNFCCC (1997), own calculations.

Participants The project developer is Reforest the Tropics Inc., referred to by USIJI as 'a not-for-profit, non-stock organization' (USIJI, 1996, p.68). It is a subsidiary of the Connecticut based forestry enterprise Newton Treviso Corporation (IUEP, 1995) and also established the Macadamia nut as a cash crop in Costa Rica some years ago (WBCSD, 1997). For 29 years it has been operating a model plantation of Klinki in Turrialba, east of San José. The Cantonal Agricultural Center of Turrialba (CACTU) will manage and monitor the project. The CACTU finances itself by selling utility poles (ibid.). Its board of directors is constituted by representatives of farmers, local banks and cooperatives (USIJI, 1996, p.68). Yale School of Forestry and Environmental Studies, The Forest Products Laboratory, which is a department of the US Agriculture Ministry and its Costa Rican counterpart, the *Centro Agronómico Tropical de Investigación y Enseñanza* (CATIE) will collaborate in the project. Last but not least, a survey among the farmers resulted in 40 farmers declaring themselves willing to plant Klinki trees on 2750 hectares (ibid., p.69). In 1998, no more than five farmers had signed planting contracts (USEPA, 1998). The Costa Rican government supports the measure, although the adverb 'strongly' is left out in this particular case (USIJI, 1996, p.71).

Climate effects In the first year, only 100 hectares are to be planted because the Costa Rican production capacity of the seedlings is very limited. Annual carbon fixation per hectare is said to be 8.2 tons, referring to the model plantation. The growth is consciously overestimated for the first years and underestimated for later years. This simplification is not acceptable, because the project developer could well offer exact data for the first 30 years from the Turrialba model plantation of Klinki pines. After an even more simplistic calculation model in the first report to the FCCC (USIJI, 1996, p.73), the project developers adapted carbon calculation, time schedule and costs. Carbon calculation now takes the results of first year plantations into account in the second year, and so on. As cost estimations nearly tripled, project lifetime was extended. In the last five years carbon sequestration slows down as a result of massive cutting. Year 45 sees the 6000ha pine forest reduced to 1561ha (UNFCCC, 1997).[11]

Carbon profits are not balanced against losses, because the reference case is zero. The carbon content of pastures is supposed to be stable, and carbon emissions owing to the cutting, processing and consumption of the wood are not taken into account.

Costs The developers expect project costs of US$10 666 017, 30 per cent of which are transaction costs such as inventory (9.1 per cent), monitoring (8.6 per cent) and project management (11.8 per cent). The derived price is US$5.42 per ton of mitigated carbon.

Externalities The implementation of new, non-traditional agrarian products that help to reduce dependence on coffee and banana production certainly is beneficial for the Costa Rican economy, as are increased incentives to forestry and against cattle growing. Still there is no estimate of viability for the farmers. Project developers expect net revenues from the sale of wood to exceed US$150 million (UNFCCC, 1998a). Commodity prices for wood are so low that sawmills can afford to waste 50 per cent of the raw material during processing (Butterfield, 1994, p.318). On the other hand, project developers list among the social and cultural impacts of the project that it will provide more affordable construction materials (UNFCCC, 1998a).

Most of all, there are real doubts about the Klinki forestry from the ecological point of view. There is no information given about the soils needed for Klinki, or about the water demand of the plants. There is no information about the water absorption capacity of the soils cultivated with the pine, or about how it reacts to forest fire. It is hoped, however, that it improves water habitats and streamflow (ibid.). The sole fact that Klinki can be grown in mixed species plantings is praised as 'contributing to biodiversity and

11. This number is obtained by dividing the projected carbon sequestration in year 46 by 8.2 tons.

plantation stability' (WBCSD, 1997). This is precisely not the case with the exotic Klinki species, which is to be widely planted. The 30 years of experience in the model planting do not even account for one generation of trees. It is quite probable that they will be subject to infestation and will themselves damage other organisms within the tropical ecosystem by the secretion of needles, resin or seeds. The rapid growth will deplete the poor soils of the pastures. The use of pesticides and fertilizers will be indispensable. Contrarily, project developers argue that the plantations would 'reduce the use of chemicals as low-yield crops are shifted to forest plantations' (UNFCCC, 1998a). This again sheds some doubt on the project developers' definition of pastures. An uncontrolled spreading into the ACCVC buffer area adjacent to Turrialba could lead to unpredictable consequences. As no new plantings are projected in the later years, KLINKIFIX may leave 60 square km of fallow land behind.

Observations In the Klinki case, the criterion of programme additionality is not fulfilled. On the contrary, the programme was planned and prepared for three decades before it was submitted for AIJ approval. The task of AIJ is to encourage measures that otherwise would not have taken place. Klinki developer Reforest the Tropics, Inc. freely compares the programme to its previous experience with the implementation of the Macadamia in Costa Rica, which also did not require external financing. An investment which would have taken place anyway is expected to be 'sweetened' by the expected climate benefits. In our opinion, Newton Treviso Corp., alias Reforest the Tropics, Inc., has to be regarded as the typical free rider in the AIJ process. The company is the only provider of the seedlings, and contractor of the farmers for planting and for the marketing of their carbon credits. Only the commercial risk of selling the wood harvest is left to the farmers.

CACTU is too closely involved in the project outcomes to be suitable for independent monitoring. In March 1998, project director Herster Barres declared his expectation 'that the sequestered carbon ... will be certified by the Government of Costa Rica' (Barres, 1998). As the project obviously lacks investors, he now goes fund raising in schools and churches for 'pre-project implementation', as the uniform reporting format (UNFCCC, 1998a) states. Nevertheless, KLINKIFIX is reported to have started in June 1997.

Obviously, those are the reasons why the Costa Rican government gives only its lukewarm support. In a personal conversation, Franz Tattenbach admitted that he did not feel too sure about KLINKIFIX, but the project developer had been so enthusiastic about it, OCIC did not want to disappoint him. Perhaps owing to the fact that two national bodies have to decide on the applications, the job of refusal is mutually left to the other side.

Virilla river basin project

The Virilla project is special in many ways. It is a combined energy *and* forestry project. However, crediting is given only for the forestry part. It is financed by the Norwegian foreign ministry. The project is located 20km north-west of San José, near the airport and the free-trade zone. There is an older 1MW power plant which will be replaced by the new 28MW generators. The waters are so heavily contaminated by the industry of the nearby free-trade zone, that a Norwegian technical manager preferred to call the project a 'sewer power plant' (Anonymous, 1997a). A primary forest area of 2000ha. is being put under protection. A secondary forest of 1000ha will be managed and protected, with an annual incorporation of 100ha, so that the total area will be reached by the year 2006 (see Table 6.3). Another 1000ha will be reforested, starting with the year 1999, in annual steps of 100 hectares. The protection zone covers the upper Virilla river basin, limited to a forest reserve and a national park. Implementation will take 10 years, but quantification and monitoring will go on for 25 years. However, the mitigation is guaranteed to the investor for no more than 20 years. Certificates will be issued during the first five years, so that the last certificate covers the period from year 6 to year 25.[12] The forestry part of the project is financially administered by the National Forestry Fund, FONAFIFO. It is part of the Private Forestry Project (PFP) mentioned above, which provides compensation for sustainable forestry.

Participants The Costa Rican side is represented by the *Compañía Nacional de Fuerza y Luz* (CNFL), the power production branch of the *Instituto Costaricense de Electricidad* (ICE). The lion's share of US$1.7 million comes from the Norwegian Foreign Ministry. These funds are provided by the Norwegian Climate Fund and will not be reported as part of the official development aid (Jepma, 1997, p.18). The *Consorcio Noruego* (CN) is a consortium put together for the occasion by ABB Kraft and Kværner Energy, Statkraft Engineering, Atlas Copco, Ølsen Stålindustri and Norman Olsen Maskin. CN spends US$300 000 on mitigation certificates and is compensated by a purchase commitment for 65 per cent of all goods and services needed for the power plant to be provided by Norwegian companies (Anonymous, 1997a). Their financing will be facilitated by official Norwegian export credits. The funds are distributed through the PFP. Over 900 landowners were initially expected to participate. In the Virilla area the PFP has not been as successful as in other parts of the country. The reason is that the programme is more attractive to beef-producing peasants than to

12. See below for the Costa Rican mitigation certificates called 'Certifiable/Certified Tradable Offsets'.

dairy farmers who have a wider margin of profitability, and land prices are quite high. As of mid-1998, only 12–30 landowners were reported to have signed the FONAFIFO agreement (Subak, 1998, p.8).

Climate effects Carbon credits are given only for the conservation and reforestation part of the Virilla project (see Figure 6.1). An average estimated deforestation rate of 7.5 per cent seems overstated, because the overall deforestation rate in the same time period is about 3.2 per cent and deforestation is constantly declining. The estimate results from Landsat image evaluation for the years 1986–92, confirming theoretical findings by FUNDECOR. Perhaps because of the proximity to the capital, deforestation pressures may be higher than average. Carbon contents per hectare in this case are estimated at only 67 tons. Compared with the ECOLAND project which calculates 110t/ha in biomass (not accounting for soil fixation), in the field of biomass fixation Virilla is very conservative. Thus the two potential baseline estimation errors in this case may rule each other out.

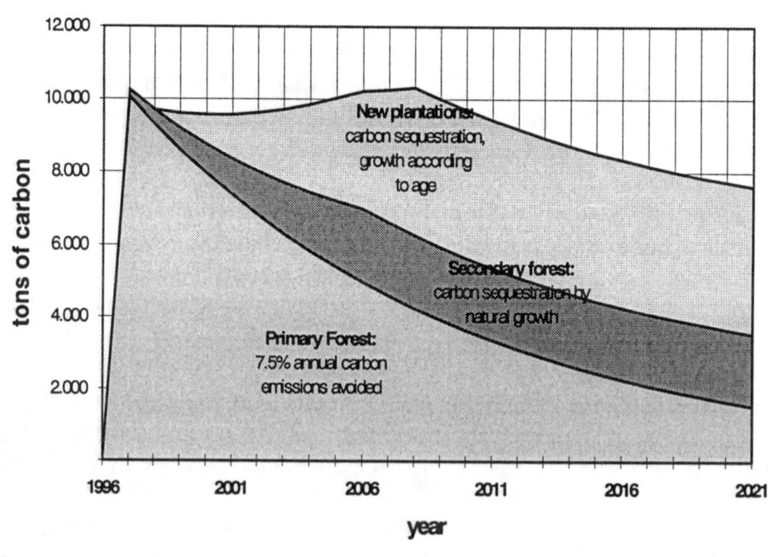

Figure 6.1 Virilla basin project mitigation effects

Table 6.3 Virilla: carbon stored in forest and fixed by plantations

	Primary forest [a]		Secondary forest [b]		Plantations [c]		'Cumulative effect'
	Deforestation (ha)	Carbon emissions avoided (t)	Accumulated hectares	Carbon sequest-ration (t)	Accumulated hectares	Carbon sequest-ration (t)	Total carbon with project (t)
1996	2 000	(-)	(-)	(-)	(-)	(-)	(-)
1997	1 850	10 050	100	200	100	(-)	10 250
1998	1 711	9 296	200	400	200	(-)	9 696
1999	1 583	8 599	300	600	300	405	9 604
2000	1 464	7 954	400	800	400	810	9 564
2001	1 354	7 358	500	1 000	500	1 215	9 573
2002	1 253	6 806	600	1 200	600	1 620	9 626
2003	1 159	6 295	700	1 400	700	2 025	9 720
2004	1 072	5 823	800	1 600	800	2 430	9 853
2005	992	5 386	900	1 800	900	2 835	10 021
2006	917	4 982	1 000	2 000	1 000	3 240	10 222
2007	848	4 609	1 000	2 000	1 000	3 645	10 254
2008	785	4 263	1 000	2 000	1 000	4 050	10 313
2009	726	3 943	1 000	2 000	1 000	4 050	9 993
2010	671	3 648	1 000	2 000	1 000	4 050	9 698
2011	621	3 374	1 000	2 000	1 000	4 050	9 424
2012	575	3 121	1 000	2 000	1 000	4 050	9 171
2013	531	2 887	1 000	2 000	1 000	4 050	8 937
2014	492	2 670	1 000	2 000	1 000	4 050	8 720
2015	455	2 470	1 000	2 000	1 000	4 050	8 520
2016	421	2 285	1 000	2 000	1 000	4 050	8 335
2017	389	2 113	1 000	2 000	1 000	4 050	8 163
2018	360	1 955	1 000	2 000	1 000	4 050	8 005
2019	333	1 808	1 000	2 000	1 000	4 050	7 858
2020	308	1 673	1 000	2 000	1 000	4 050	7 723
2021	285	1 547	1 000	2 000	1 000	4 050	7 597
Total		114 917		41 000		74 925	230 842

Notes:
[a] Deforestation rate: 7.5%; Carbon content per ha: 67t C.
[b] Annual growth of carbon stock: 2t C/ha.
[c] Annual growth of carbon stock: 4t C/ha.

Sources: Jepma (1997, p.17), UNFCCC (1997), own modifications.

The formula for calculating the percentage of forest lost (L) is

$$L_n = 1 - (1 - p)^n.$$

If the observed annual deforestation of 7.5 per cent were to continue, over 25 years ($1 - (1 - 0.075)^{25} =$) 85.7 per cent of the forested area would be lost. The projected total carbon stock reaches 230 842t in 25 years.

Costs The project costs are cited by the proponents as US$3 395 243 (UNFCCC, 1998a). The CNFL will invest US$1.39 million. This is the amount the utility would have spent for forestry issues without the project. Mitigation certificates are generated for the investment of the Foreign Ministry and of *Consorcio Noruego*. Relating carbon benefits from the forestry part of the project to total *external* financing results in US$8.66 per ton, not taking into account operational returns during the project phase. Including the CNFL contribution, the price is US$14.71 per ton (see Witthoeft-Muehlmann, 1998, p.24). On the other hand, these costs seem underestimated because the PFP backs financing for forestry subsidies. Landowners receive payments over five years. Their contracts commit them to ten years for forest management and only five years in the case of plantations (Subak, 1998, p.11). If they disregard the contract, they have to pay back the total payment received. Depending on the future development of interest and inflation rates, the opportunity costs for these lands might be higher than the choice of paying back the subsidies and clearing the lands. Subak (ibid., p.24) finds the penalty system 'very weak'. She estimates that the real cost 'is closer to 30$/t in protected forests and 40$/t for carbon sequestered in plantations' (ibid., p.40).

Externalities Forest and watershed conservation contribute to the improvement of the deteriorated environment of the San José region as well as to the improvement of soil and water quality and the hydrological regime of the area. Both will lead to an increase in energy production, estimated at between 7 per cent and 9 per cent (UNFCCC, 1997). The power plant itself would not have depended directly on forest protection. Given the Norwegian know-how in hydroelectric technology, the Virilla plant may now be more durable and potent than it would have been without the cooperation. There is an external profit from the watershed protection for the hydroelectric plant resulting from the extended availability of water in the dry season.

Until mid-1998, landowners' participation was very low. Subak counted between 12 and 30 participants (1998, p.8). She concludes, that the forestry subsidies are attractive to beef-producing farmers, but not to dairy farmers in an area where land values are relatively high.

Observations The beginning of the works was planned for November 1996, but it was delayed owing to problems with financing and contracts until March 1997. Another serious problem was red tape: as no heavy machinery could be obtained in Costa Rica, it had to be brought into the country. Customs declaration was highly complicated and costly. Coordination manager Rolf Thorsen complains bitterly about Costa Rican bureaucracy (Anonymous, 1997a), thus contradicting the pretensions of the OCIC. Technical problems were posed by the notorious lack of infrastructure and the fact that the 69m high dam was to be built in an earthquake-prone area.

As far as the forestry part is concerned, it seems to be more or less within the plan. Until mid-1998, about 25 per cent of the planned forest had been included in the project area (Subak, 1998, p.8). However, owing to a certain resistance among the landowners, it is doubtful whether the incentives offered are high enough. For the actual participants, Subak, who investigated in the area, alleges that 'the Private Forestry Project may not be the major motivating factor behind participation' (ibid., p.21). If her assumption that they would have participated anyway because of their individual environmental preferences is true, this would question the criterion of programme additionality.

Energy Projects

Power production is dominated by the state-owned electricity and telecommunications monopoly, *Instituto Costaricense de Electricidad* (ICE) with its regional subsidiaries. Although Costa Rica has a large potential for renewable energy, the indebtedness of ICE and legal constraints against private generation hinder the power grid from keeping pace with the steadily growing demand (LeBlanc, 1997, p.3). Privately owned renewable energy facilities are limited to 20MW and are not allowed to constitute more than 15 per cent of total system capacity. Another maximum 15 per cent can be covered by tender for no more than 20 years. Only companies with at least 35 per cent of national participation are allowed to compete, and the cumulated limit for each provider is 50MW. After the period has run out, ownership of the plant is transferred to the ICE. This procedure is known as the build–operate–transfer (BOT) scheme (InterAm Database, 1995). A recently proposed amendment limits small private renewable energy production to 5MW for each plant and to 5 per cent of system capacity (LeBlanc, 1997, p.4). If this proposal is made law, prospects will be bad for small energy projects. However, if the new government should decide on the long discussed privatization of the ICE, these restrictions could soon be made obsolete.

Owing to the phase-out plan mentioned above, baseline projections are very restrictive for all power projects. In the USIJI report, each project has its own baseline, based on different data. In order to make emission reduction comparable, the Aeroenergía baseline was chosen, because its cumulated bias is lowest when applied to the other projects. The given data are set in italics in Table 6.4.

Table 6.4 Reference case fossil energy production

Year	Phase-out plan (%)	Fossil production (GWh)	Carbon emissions (t)
1994	*0*	*829.8*	*235 048*
1995	*0*	829.8	235 048
1996	*0*	829.8	235 048
1997	*0*	829.8	235 048
1998	*44*	464.7	131 627
1999	*86*	116.2	32 907
2000	*99*	8.3	2 350
2001	*100*	0.0	0
C emissions/GWh (t)			*283.2586*

Source: USIJI (1996, p.40), own calculation.

Neither of the projects puts out tenders for goods and services. Contracts usually include provider commitments.

Wind power

The pacific rim of Guanacaste is best suited for wind power generation. All the three projected facilities are located in the area around Tejona de Tilarán. The hilly plateau has been a cattle-farming area for centuries. Winds are extreme in the dry period between January and August, thus making wind power an ideal match for hydropower.

The *Aeroenergía* (AE) wind park is the smallest, with a capacity of 6.4MW. *Plantas Eólicas* (PE) and *Tierras Morenas* (TM) are both projected for the maximum level of 20MW. All of them are strongly supported by the Costa Rican government. Although OCIC official Adalberto Gorbitz (1997, p.53) states that all energy projects are fully financed, PE and AE are the only projects whose realization has been confirmed, although much later than originally expected.

Up to the present, home markets for wind power stations are mainly in Northern Europe and Northern America. Manufacturers are therefore interested in gaining experience under tropical conditions and gaining market entry in the developing countries. The main features of the three wind power projects are listed in Table 6.5.

Table 6.5 Features of the wind energy projects

	Aeroenergía	Plantas Eólicas	Tierras Morenas
Approval	July 1995	November 1994	July 1995
Participants	Aeroenergía S.A.,CR Energy Works (subsidiary of USA-based Bechtel Corp. Power Systems Inc., USA Bluefields International, USA Micon A/S, Denmark	Plantas Eólicas S.A. (joint venture between Merril International, USA and Charter Oak Energy, subsidiary of Northeast Utilities, both USA) Kenetech Windpower,USA	Energy Works, USA New World Power Corp., USA[13] Molinas de Viento del Arenal S.A., Costa Rica MINAE (CR Ministry for Environment and Energy)
Suppliers	Micon A/S	Kenetech Windpower	Enercon, Germany
Monitoring	Aeroenergía S.A.	Plantas Eólicas S.A.	MINAE
AnnUAL Production	30GWh	98GWh	90GWh
Starting date[14]	May 1997 (UNFCCC, 1998a)/ July 1998 (USIJI 1998)	June 1996	September 1997[15] (UNFCCC, 1998a)/July 1999 (USIJI, 1998)
Lifetime	21 years, 1 month	15 years	13 years, 11 months
Costs	US$8.85 million	US$27–30 million[16]	US$31.5 million[17]
Financing	Central American Economic Integration Bank: 75% Aeroenergía Partners: 25%	Charter Oak Energy Tejona Corp.: 65% Manuel Emilio Montero Anderson: 35%	Molinos de Viento del Arenal S.A.: 30% Commonwealth Development Corporation: 24% International Finance Corporation: 24%

13. New World Power Corp. is not listed any longer in the latest report of 1998.
14. Beginning of operation.
15. The UNFCCC uniform reporting format dates from 13 October 1998. Contrarily to the given date, the current stage is only given as 'mutually agreed' (UNFCCC, 1998a).
16. Total funding of the plant is kept confidential by the developer (UNFCCC, 1997). Gorbitz (1997, p.55) gives the figure of US$30.4 million.
17. Gorbitz (1997, p.55) gives the figure of US$27 million.

Aeroenergía	Plantas Eólicas	Tierras Morenas
		Commercial banks: 22% *Funding not assured*[18]

Sources: USIJI (1996, pp.36–88), UNFCCC (1997).

Participants See Table 6.7.

Climate effects Calculations for emission reduction achieved can be seen in Table 6.6. The Costa Rican fossil fuel phase-out plan leads to a double disadvantage for the projects: first by comparing them to the baseline, and second by discounting the reduction. The reason for this extremely conservative calculation is to prevent the reduction's being claimed by several plants. There is no computation of the cumulative effect. Each wind park is assumed to be the only *additional* plant.

Costs In contrast to the other projects, there is little sense in calculating the value per ton of carbon reduced, because – owing to the unrealistic phase-out plan for fossil fuels – only the reductions achieved by 2001 could be considered. For the three projects, had they been constructed according to their schedules, carbon prices would vary between US$400 (PE) and US$900 (AE) per ton. Of course the delays lower the emission reduction even more. This is why the projects themselves have to be economically viable in order to be realized (See Tables 6.7 and 6.8). The Tierras Morenas proponents claim that their operational costs are only US$0.015 per kilowatt hour (kWh), which would really be competitive.

18. As of October 1998 (UNFCCC, 1998a).

Table 6.6 Project cases for wind power

	Aeroenergía			Plantas Eólicas			Tierras Morenas		
Year	Annual capacity (GWh)	Total carbon emissions with AE (t)	Weighted difference (t C)	Annual capacity (GWh)	Total carbon emissions with PE (t)	Weighted difference (t C)	Annual capacity (GWh)	Total carbon emissions with TE (t)	Weighted difference (t C)
1994									
1995									
1996				98.0	207 289	27 759			
1997	15.8	230 587	4 461	98.0	207 289	27 759	76.0		
1998	27.0	123 979	4 283	98.0	103 868	15 545	76.0	110 099	12 055
1999	27.0	25 259	1 071	98.0	5 147	3 886	76.0	11 379	3 014
2000	27.0		24	98.0			76.0		
2001	27.0			98.0			76.0		
Total AE: 9 838				Total PE: 74 950			Total TM: 15 069		

Source: USIJI (1996, pp.36–88), own calculations.

Externalities Apart from environmental benefits, renewable energy production serves both development and social objectives. Resistance to the privatization of ICE can be understood in this context. According to the 1984 census, an average of 18.1 per cent of Costa Rican households were not connected to an electricity supply. In some regions, the figure was over 50 per cent (Hein *et al.*, 1994, p.134). Investment in infrastructure is needed as a prerequisite for the creation of employment.

Direct effects on the Tilarán region will be very few. Wind power stations do not require a lot of labour. Investment in property acquisitions for the wind installations will not be important.

Observations Neither of the JI bodies of the USA and Costa Rica took much trouble in the baseline calculation, maybe because the carbon reduction is not the incentive in this case. Apart from the different databases used (while always referring to the MINAE), there were several calculation errors, and they even sometimes understated the emission reduction.

First experiences with Plantas Eólicas show that the project developers underestimated the fact that Guanacaste is a strong wind area. Although the wind harvest was higher than expected, there are serious technical problems with the Kenetech generators. One of them was blown down, and all of the towers suffer from strong vibrations. Another problem related to Kenetech is that the company went into bankruptcy in April 1997 and is currently winding up its operations. At present, nobody knows who will take the place of Kenetech in the management of the project, or whether spare parts and services will be available in the future.

Again in this case there are doubts about financial additionality. On 20 December 1995, the Inter-American Development Bank (IADB) announced a US$18.7 million loan for PE, partly granted (US$7.2 million) as a normal loan and partly provided by a commercial bank, under subscription of a participation agreement with the IADB. It also mentions an adjacent 'sister plant' which would be operated by the ICE and financed both by an IADB loan and a Global Environment Facility (GEF) grant of US$3.3 million (IADB, 1995). As IADB spokesman Daniel Drosdoff stated in September 1997, the loan was actually never signed, but the project was financed by Charter Oak. As the application to the USIJI was placed in November 1994, applicants evidently followed a double-track strategy.

The Aeroenergía project was reported to have started operation in May 1997. This fact was not known by Paulo Manso, the OCIC consultant specializing in renewable energies, in July 1997.[19] The latest USIJI report (USIJI, 1998) states that the starting date was July 1998.

Finally, Tierras Morenas no longer appears to be an AIJ project. It is mainly financed by the ICE subsidiary Molinos de Viento del Arenal and two multilateral development corporations.[20] Perhaps project developers decided to stop waiting for AIJ investors and took advantage of existing alternatives for funding. Once this funding is assured, Tierras Morenas should be removed from the FCCC listing.

Water energy

The only US hydroelectric project was approved in July 1995. The 16MW plant, Doña Julia is to be constructed in Horquetas de Sarapiquí, in the Heredia region, which has reportedly up to 8000mm of annual precipitation. The plant will be located on the rivers Puerto Viejo and Quebrandón. A reservoir will provide for peak load.

Participants The project (see Table 6.7) was developed jointly by MINAE and the New World Power Corp. Participants agreed on monitoring on a regular basis and cooperation with the ICE for the development of offset information.

---

19. This statement refers to a private conversation held on 11 July 1997.
20. According to the UNFCCC report, this funding is not yet assured (UNFCCC, 1997).

Climate effect The reported baseline estimate was replaced by the one for Aeroenergía again. Doña Julia was originally planned to come on-line in October 1996, but it will not be able to really claim the carbon credits calculated unless an investor has been found. There is no evidence that the time schedule has been fulfilled and the plant is really operational. The second report to the FCCC in October 1997 appears to have been elaborated on the basis of the 1996 report. It is still unclear whether the plant has started operations.

Table 6.7 Project case for Doña Julia

Year	Annual capacity (GWh)	Total carbon emissions with AE (t)	Weighted difference (t C)
1994			
1995			
1996	22.5	228 675	6 373
1997	90.0	209 555	25 493
1998	90.0	106 134	14 276
1999	90.0	7 413	3 569
2000	90.0		
2001	90.0		
		Total	49 712
		USIJI data	57 400

Source: USIJI (1996, pp.60f) own calculations.

Costs Estimated total costs are US$28 million. This amount is not broken down into its factors. 'The full capitalized costs of the project have been financed with a combination of debt (70%) and equity (30%)' (UNFCCC; 1997). Neither report names a creditor and/or investor.

Externalities An Environmental Impact Study was carried out by MINAE in April 1994 (USIJI, 1996, p.59). The second report to the FCCC says:

> The negative effects include obstacles to fish migration, disruption of natural ecosystems by changing river regimes, changes in land use via construction of roads and transmission lines,

> sedimentation upstream of the dam, and river bed erosion downstream. Positive effects include mitigation of floods, creation of new habitats for some animal species, recreation, and the improvement of transport in rural areas by new roads (UNFCCC, 1997).

The project will provide employment during the construction and operational phase. Nevertheless, it seems difficult to quantify this effect.

Observations USIJI admits there was no real programme additionality in the Doña Julia case, because it was conceived several years earlier. An exception was made in order to get a languishing project off the ground: 'In these cases, it should be shown that USIJI was instrumental in overcoming barriers that would, otherwise, have prevented the implementation of the project' (USIJI, 1996, p.13). The mere approval as an AIJ project is not enough as an incentive, if investors do not even see the perspective of getting credits for it. However, the uniform reporting format of May 1998 on the climate secretariat's website (UNFCCC, 1998a) claims that the power station has been operational since October 1996. In contradiction to that, a project description handed out at the Buenos Aires conference envisaged the planned starting date of November 1998 (USIJI, 1998, p.6). This means that, under the current baseline projections, practically no emission credits can be accounted for.

Climate cooperation in waste treatment

Astonishingly enough, up to the present no industrial project has been proposed for AIJ cooperation in Costa Rica. Only recently, one agroindustrial AIJ project between Costa Rica and the Netherlands has been disclosed (Anonymous, 1998; UNFCCC, 1998a). Taking as its starting-point the problem of waste water treatment in coffee processing (as depicted above), it finances additional investment for greenhouse gas mitigation. Since 1990, coffee mills have been obliged to clear their waste waters in so-called 'open lagoons'. The disadvantage of this is major methane emissions. These emissions can be avoided by a closed, anaerobic reactor producing biogas. Four of these reactors have been constructed in different coffee mills. The projects run for 10 years, equal to the economic life-span of the equipment. The implementation was between 15 August 1997 and 31 March 1998.

Participants The Costa Rican Coffee Institute realized the feasibility study. Four local coffee mills (Cooperativas Naranjo, Jorco, Libertad and Cafetalera Pilas) will implement the technology in their plants. Biomass Technology Group BV (BTG) of the Netherlands developed the reactor process and Swiss-dominated Amanco de Costa Rica S.A. received the BTG know-how to market and install the reactors. The project is financed by the Dutch government.

Climate effects Over the project's lifetime, about 5000 tons of methane emissions will be saved in all the mills taken together, compared with the baseline scenario of using open sewage lagoons. This equals 126 780 tons of CO_2 or 34 545 tons of carbon equivalents using the IPCC conversion factor.

Costs For the coffee producers, there would be a need to adapt their water treatment anyway to comply with an agreement with the MINAE signed in 1992. This agreement, however, allows open lagoons. In this example, the incremental costs for implementing mitigation are easily defined by the difference between the two alternative systems (see Table 6.8). The concept of incremental costs makes the coffee mills project rather cheap. There is no estimation given on the gains generated by the use of methane for energy use. The reason is that the mills mainly use firewood for that purpose. Another point seems to be the short project life. The commercial lifetime of the reactors will certainly be higher.

Table 6.8 Costs of methane mitigation in coffee mills

	Naranjo	Jorco	Libertad	Cafet. Pilas	Total
Anaerobic reactor system costs	342 000	117 993	313 170	159 000	932 163
Alternative treatment costs	211 402	70 325	185 845	92 334	559 906
AIJ component costs	130 598	47 668	127 325	66 666	372 257
Saved GHG emissions (t C)	13 436	6 324	11 011	3 774	34 545
GHG abatement costs ($/t C)	9.72	7.54	11.56	17.67	10.78

Source: Anonymous (1998, p.13), own calculations.

Externalities The project is the topping of a business-as-usual project for the protection of water resources. Therefore the regional environmental effects are not additional, apart from the replacement of firewood. Awareness of biogas use may perhaps increase. Labour effects were positive during the construction phase. A new technology especially suitable for coffee production has been developed and transferred. Bad odours were reduced and the neighbourhood suffers less from insects. Negative influences such as bad odours and methane escape now occur only when the plants are malfunctioning (UNFCCC, 1998a).

Observations The Dutch JI Registration Centre together with Costa Rican OCIC issues certificates. Two independent institutions from the Netherlands and Costa Rica developed the baseline scenario. The source does not give the name of these institutions, but it says that 'a first monitoring study is scheduled for early 1999' (ibid., p.13). It is interesting to note that investor and host country share the certificates accrued. If the project is to be transferred to the CDM phase, where credits will be accounted against real emissions, the question will arise as to what Costa Rica will do with them. Will it try to participate in emissions trading (which is actually not envisioned in the Kyoto Protocol) or will it claim the selling proceeds from the Netherlands?

DEVELOPING THE INSTRUMENT OF COOPERATION

From the very beginning, Costa Rica has handled AIJ in a very autonomous way. The first projects stemming from cooperation with USIJI are being financed or are still seeking funding on a project-to-project basis. The investors are directly involved in the proposal, planning and implementation of projects. For these projects the transaction costs are high; return on investment as well as the real GHG effects are submitted to considerable entrepreneurial risk. This is why Costa Rica has developed an alternative model of financing climate cooperation.

The invention of so-called 'certifiable tradable offsets'[21] (CTOs) was the unilateral anticipation of an international crediting system. Costa Rica thus envisaged a CDM-type system very early where it could sell the CTOs on an international market. Each CTO stands for an amount of GHG reduced or sequestered in vegetation, expressed in carbon equivalents.[22] MINAE guarantees the amount to the CTO buyer for the period of 20 years. This means that, if any of the financed projects fails or does not produce the expected GHG effects, the state of Costa Rica will provide for other projects to have the same effect. For instance, if CTOs are sold over five years, the effective project duration is 25 years to guarantee the 20 year existence of greenhouse gas offsets to the last buyer. This is done by selling only a part of

21. Later renamed 'certified tradable offsets', after a verification system was in place.
22. One ton of carbon equals 3.67 tons of CO_2.

all possible carbon reductions as CTOs. Each single CTO (equivalent to one ton of carbon) was initially sold for US$10, while later CTOs are offered for US$20. As soon as emissions trading is begun, the price will be derived on stock markets. The whole mechanism depends on two conditions before the start of the CDM: (a) the home country of the buyer will recognize CTOs as valid, and (b) CTOs are of economic use to their holder (for example, entitles him or her to tax exemptions). Under the CDM, it is still unclear whether host countries may freely trade credits.

Proceeds from selling CTOs go to the above mentioned National Forestry Finance Fund, FONAFIFO[23], which is responsible for the so-called 'umbrella projects'. In theory, these projects only need the approval of the Costa Rican JI office, and not of the investor's country. On the contrary, CTOs would not be transferable. In practice, the state of Norway as the first buyer of CTOs developed two projects together with OCIC. The overall project volume being US$3.4 million, Norway obtained 200 000 CTOs for the first US$2 million invested in forest conservation and reforestation.

What are the advantages of CTOs over the project-to-project approach? Investors can enter cooperation on a very small budget because the virtual minimum investment could be US$10 for one CTO. No administration is required and there are no external costs. The investor takes no risk if one project fails or is delayed; the only risk lies is the stability of the constitutional system of Costa Rica, which in turn has proved stable over the last 50 years. On the other hand, there is less free-riding for the investor by linking the involvement to supply contracts. This conforms with the restrictions set by the UN Framework Convention. On the other hand, each CTO is 'tagged', meaning that it is linked to a concrete project which can be traced at any time. This is to prevent mitigations being sold twice.

The benefits for Costa Rica consist in realizing projects according to its own economic necessities and political preferences, thus fully conserving its sovereignty. This is reflected in the decision of OCIC to withdraw projects that have not yet received financing on the project-to-project basis and to carry them out under the umbrella of the Forestry Finance Fund.

Because CTOs are insured against failure, they offer the opportunity to Costa Rica to sell greenhouse gas offsets that will be achieved in the future.

---

23. In other publications reference is made to a 'National Carbon Fund' (Foundation JIN, 1996, p.2). This fund is not clearly defined. It seems to be just a part of FONAFIFO.

CTO-financed Umbrella Projects

There are actually two 'real' umbrella projects being financed by fuel tax and CTO selling.[24] Besides the PAP, which primarily serves the national park administration for purchasing land, there is the Private Forestry Project (PFP) which 'aims to compensate farmers for forest conservation, reforestation or sustainable management efforts' (Foundation JIN, 1996, p.2). Whereas each CTO (equalling one ton of carbon mitigated) was initially (in the Virilla case) sold for US$10, after the SGS certification the OCIC raised this price to US$20. The reforestation part of the Virilla project forms part of the PFP. The PFP has no clearly defined boundaries and can therefore be described as a programme rather than a project. The second unknown factor is the magnitude of internal finance within the PFP (fuel tax). Furthermore, no time-frame has been given. In the author's opinion, much work remains to be done to prove financial and project additionality of the PFP.

In July 1997, the US Initiative on Joint Implementation (USIJI) bought 16 million CTOs, 11 million of which are to be achieved by forest conservation and the rest by reforestation. An independent verification of all CTO-related projects by the Swiss company SGS has been agreed upon, which is financed by the World Bank. Up to the present, there is no private holder of CTOs, except for the USA-based company Environmental Financial Products,[25] which bought a 1000-ton CTO at a price not revealed to the public (Liddell and Escofet, 1997, p.12) and placed it on the Chicago stock market.

Protected areas project

The Protected Areas Project (PAP, *Proyecto de Áreas Protegidas* in Spanish) comprises 'paper parks' which were declared national parks or forest reserves but do not belong to the state (see Figure 6.2). It is an attempt to join forces on a CTO base. Its surface totals 530 498 ha, no less than 41.9 per cent of all Costa Rican areas under any kind of protection. As stated above, it bundles the former BIODIVERSIFIX and CARFIX projects, which have not been reported since July 1998. The proponents declare explicitly: 'this project will not claim GHG benefits for any areas that are accruing GHG benefits as a result of their involvement in other active AIJ projects' (UNFCCC, 1998a). This refers especially to the Piedras Blancas national park. As a part of its surface is subject to the ECOLAND project, it only accounts for 11 289 ha

24. Another CTO programme, the Costa Rican Renewable Energy Export Programme, which totals 215MW of capacity to offset 0.35 million tons of carbon per year in the importing neighbouring countries (LeBlanc, 1997, pp.12f) seems to be languishing at present. There are technical shortcomings and incompatibilities of the electric grids between the different Central American states which question the programme's feasibility.
25. Renamed in 1997, formerly Centre Financial Products.

within the PAP.[26] Furthermore, an Earth Centre will be built, 'a multidisciplinary development combining residential, commerce, and work activities to provide public education and entertainment and to promote ecotourism' (ibid.).

Participants The National Parks Foundation and the Costa Rican Earth Council Foundation cooperate on the Costa Rican side. They are supported by the US counterpart of the Earth Council. The bulk of the funding is expected to result from the sale of CTOs offered by the Chicago-based Environmental Financial Products Ltd.[27]

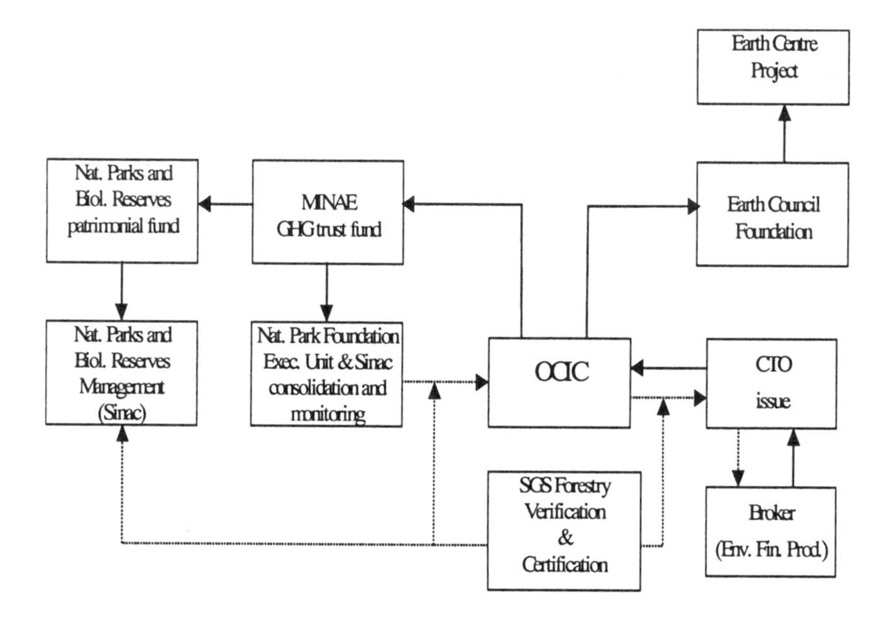

Source: SGS (1998, p.17).

Figure 6.2 Implementation of the protected areas project

Climate effects Constant monitoring and certification of the project is assured by the Swiss Société Générale de Surveillance (SGS Forestry). Its report (SGS, 1998) has been made publicly available. SGS will certify on an annual basis the mitigation effects actually achieved. After scrutinizing the different subprojects thoroughly, SGS set preconditions for certification,

26. See the ECOLAND section above.
27. In the July 1998 report, the old name and address of the company are still reported.

among which is the obligation to set up a buffer. Because of the high risk of forest losses owing to erosion and fires in some regions, only some 56 per cent of the potential offsets of 1.76 million tons of carbon can be sold after they have accrued. The remaining offsets have to be set aside, at least until the uncertainty is reduced. CTOs are now called 'certified tradable offsets'. The now included BIODIVERSIFIX and CARFIX projects have been completely recalculated. The SGS baseline assumptions are very conservative, especially concerning the deforestation rate. SGS observed two trends in opposed directions: 'Deforestation as a result of the rising demand for agricultural products is not expected to continue in the long term', and 'Population growth and increasing consumption of timber for construction and export are likely to maintain a proportion of deforestation trends in the future' (ibid., p.29).

The average deforestation rate within protected areas based on satellite imagery was estimated to be 1.47 per cent. SGS uses a rate of 1.08 per cent for the baseline calculation. 'Another 16% to cover the uncertainty related to the positioning of the baseline' (ibid.).

Costs The total costs are estimated at US$158 138 898, of which US$20 000 000 will go to the planned Earth Centre. The certification and handling overhead, US$1 033 735, is expected to be financed by the CTO buyers. This aspect is questionable, however. Mulongoy *et al.* (1998, p.21) calculate sequestration costs for Costa Rica at about US$24 per ton of carbon, which correlates to the estimate given by Subak.[28] This makes the proceeds from CTO sales only *one* contribution to the governmental programme, the other being the above-mentioned proceeds from the fuel tax. On the other hand, if the fuel tax really were abolished, mitigation could no longer be offered at the same price.

Externalities The project is part of the Costa Rican development strategy which fosters education, ecotourism, bio-prospection and the preservation of natural resources. The non-inclusion of Indian reserves around La Amistad National Park near the Panamanian border is explicitly given mention.

Observations The PAP is seen by the Costa Rican government as the third stage of climate cooperation. In fact, it seems the most mature of the projects presented in this chapter. However, in the light of the current negotiations, it is questionable whether the PAP can still be called *one* climate project. It is, rather, a government programme partially funded by foreign investors.

28. See the Virilla case, above.

This leads to some doubts as to financial additionality. If climate cooperation funding only covers part of the real costs, the resulting credits cannot be fully attributed to the foreign investment. The way it is calculated today would have to be regarded as a subsidy for carbon sequestration services, or even unfair competition against other participating countries. Furthermore, it does not comply with the Article 6 (5c) of the Kyoto Protocol, demanding 'reductions in emissions that are additional to any that would occur in the absence of the certified project activity'. In the case of the former CARFIX project which now has been included in the PAP, USAID development financing was even withdrawn in order to facilitate its inclusion in the AIJ programme (Tattenbach, 1996). This is opposed to the purpose of climate cooperation. In future, a *modus vivendi* will have to be found between development aid and GHG mitigation investment.

CONCLUSIONS

Costa Rica offers far better institutional and political conditions for climate cooperation than the vast majority of developing countries. This makes the small Central American country an appropriate testing ground for the viability of this flexible instrument. The key factor of the Costa Rican AIJ policy is that the process is host country-driven (Gorbitz, 1997, p.56). Climate cooperation requires reliability and longevity of the institutions involved, making it a mechanism not applicable to least developed countries in which extreme poverty and political risks prevail.

Still, the mechanism itself exhibits shortcomings which ought to be mentioned. They can be found in baseline scenarios, financial additionality, suitable project forms, 'no-regrets', and the procedure of approval.

Conflicting Baselines

It is particularly interesting that the forestry projects try to define baselines for both afforestation and the preservation of existing stocks. With one exception (ECOLAND), they do not account for sequestration/emission of the soil. In Costa Rican projects, the extrapolated deforestation rate is taken as the baseline (USIJI, 1996, p.65). If there are uncertainties about existing stocks on deforested lands which can easily reach a considerable order of the magnitude, ECOLAND chooses the upper bound as the baseline (ibid., p.47) while KLINKI sets the stock after deforestation to zero (ibid., p.54).

Renewable project baselines do not include life-cycle emissions of the plant material. A very interesting aspect of the renewable energy projects in Costa Rica is that, because of the commitment of the Costa Rican government to phase out fossil fuel electricity production by 2001 the baseline is zero emissions after 2001. USIJI doubts that this commitment can be fulfilled (ibid., p.39), but nevertheless requires the baselines to take it into account. This means that renewable energy projects will not become creditable under the CDM regime after 2000. Therefore all the renewable energy projects actually approved and implemented are likely to be 'no-regret' projects.

The pilot phase should be seen as a chance to gain experience with GHG calculations. There are many different approaches as far as reference and project cases are concerned. While project cases can be monitored during the realization, the baseline problem remains crucial. Usually, baselines are static and remain valid for the whole project lifetime. Only in the CARFIX case was the possibility considered of keeping the baseline dynamic by monitoring the surrounding deforestation. This proposition is a new approach to the baseline discussion which should be observed in the future.

Sometimes the JI institutions involved seem to have neglected the importance of the climate effect calculations and therefore did not even take the trouble to check them. They should consider that after the actual pilot phase, these carbon credits might entitle the holder to GHG emissions in his or her home country. There is a need to issue credits only when they actually accrue – not in advance. Emission credits for the ECOLAND investor were accounted for right away, long before the carbon benefits were achieved. In the later AIJ stages, the risks associated with the imminent payoff were covered by the CTO 'buffer', which is a kind of insurance. In future, insurance companies are likely to cover these project risks. This will act as an additional incentive for the accuracy of monitoring and verification.

Financial Additionality

As illustrated in the PAP case, the question of financial additionality remains controversial. This relates not only to unfair competition by means of internal subsidies to climate projects, but also to climate-relevant development cooperation. If climate cooperation is to be beneficial for development, too, it seems sensible that the two forms of financing should be bundled together whenever a project equally fulfils both priorities. However, the question is who should be allowed to harvest the fruits of mitigation. Clearly, greenhouse gas mitigation is a side-benefit to many development projects and need not be accredited to the investing country or international institution, as in the case of the Tierras Morenas wind park. But whenever the World Bank or national ODA institutions finance activities which enable climate capacity building in order to attract third-party investors, should these investors receive the full amount of credits once the project is in place? For instance, the SGS study on the PAP was financed by the World Bank. Can project development, administration, monitoring and verification be separated from its outcomes?

Project Forms

With the recent exception of the coffee mills project, there are only energy replacement and carbon sequestration projects in Costa Rica, although a large number of alternatives could be considered. In terms of GHG reductions, forestry projects have been preferred. There are several reasons for this:

1. Most projects were proposed by or conducted in close cooperation with the Ministry of Environment and Energy (MINAE), which is very much concerned with conserving the remaining forests and uses AIJ as one instrument of international financing among others.
2. At first glance, forestry projects offer vast carbon mitigation benefits, when comparing reduction, sequestration and non-emission. However, if baselines and risks are assessed in a realistic way, the real costs may become similar. In the future CDM regime, preference could be given to technological reductions which cannot be reversed by attributing a higher amount of credits to these options.
3. A political declaration – the phase-out plan for fossil energy – was taken as the baseline for the energy projects, thus negating the good intention by obstructing investment.

This last reason points out the principal problem of contradictory incentives for governments. If environmental regulations are too strong they tend to spoil the baseline for climate projects. The dilemma is between the most favourable baseline calculation for a country whose environmental policy is worst and the selling out of each and every environment-related measure in the name of greenhouse gas mitigation. This last possibility would conflict heavily with the additionality criterion.

No-regret Criterion

There are a lot of emission reduction opportunities which are profitable either for a company or for a country as a whole. The latter include externalities such as the reduction of other pollutants. Now the question arises whether these so-called micro- or macroeconomic 'no-regret' projects are included in the baseline. So far, the question of ' no-regret' opportunities has led to heated debates among economists. While some say that there can be no ' no-regret' projects as such opportunities would have been grasped immediately (for example, Sutherland, 1996), others estimate that 10–30 per cent of today's emissions could be reduced via ‘ no-regret’ projects (IPCC, 1996).

These differences come from the fact that, despite the theoretical profitability of many options, there are regulatory and judicial obstacles, a lack of information and skilled personnel as well as organizational rigidities. It is often reported that managers do not invest in raising energy efficiency even if its internal rate of return is much higher than the prevailing market interest rate. The main reasons are probably short planning periods, the requirement of a minimal rate of return much higher than market interest rates and a lack of capital. It is not surprising that private households have even higher thresholds for internal rates of return. This applies particularly to countries in transition (Nordic Council of Ministers, 1996, p.28) and developing countries. Often an investor cannot realize a gain as it is an externality accruing to others. Therefore, it seems that pure microeconomic ‘no-regret’ opportunities are rather scarce, whereas macroeconomic ‘no-regret’ abounds.

In practice, programme and financial additionalities can hardly be separated from one another. The objectives of MINAE have been realized in different project forms, none of which was induced by AIJ cooperation. Some, perhaps most, of the projects had been developed before and were proposed under the AIJ regime. If this were not the case, the chances of realization and subsistence would be considerably lower. The same applies to financing: from the host country's perspective, AIJ is simply another opportunity to raise funds for projects that bear positive externalities. If the rules for the AIJ pilot phase had been observed strictly, many Costa Rican projects would have been disapproved because the proponents showed too much involvement and consequently might have found other ways of raising funds. In two cases financing was withdrawn in order to clear the way for approval as AIJ projects. The rational alternative would have been to pool the money and to enlarge the projects. This is how CTOs were created. They guarantee each investor a certain amount of carbon offsets, without interfering in the rights of any other investor. These ‘carbon bonds’ need to be confirmed by an international clearing house if a carbon trading system is to be established in the future.

A critical point is the lack of information. In the author's opinion, a project should not be approved if the developers hold back relevant data. This concerns the source and amount of funding (as in the case of Plantas Eólicas), the baseline and project case calculations, ecological and local economic side-effects and post-project prospects (KLINKIFIX).

Procedure of Approval

It is hard to understand why two national JI bodies need to approve each project applying the same set of criteria. It could be demonstrated that the double approval did not guarantee an efficient control. As both bodies are interested in carrying out the programme, a kind of complicity may even arise. Supposing the guest country is interested in climate effects, while the host country is interested in positive externalities, each of them could confirm the relevant components.

Nearly all reports to the FCCC try to give the impression that the project is currently in progress, even if funding is not assured. Monitoring and verification by independent institutions will be pivotal for credibility in the future.

REFERENCES

Alfaro, Marielos (1997), 'Project Carfix in Costa Rica', in Kalipada Chatterjee (ed.), *Activities Implemented Jointly to Mitigate Climate Change. Developing Countries' Perspective*, pp.197–212, New Delhi.

Alvarado Davila, Royden (1997), 'Pretenden minar impacto bananero en medio ambiente", *La Nación*, ed. Electrónica, 19 March.

Anonymous (1992), 'Costa Rica weist Deutschen wegen Umweltzerstörung aus', *Frankfurter Rundschau*, Frankfurt/M (92), 21 September.

Anonymous (1993), 'En Menos de Tres Décadas Podrían Agotarse los Bosques Costaricenses', *El Financiero*, Mexico DF, 21 January, p.31.

Anonymous (1997a), 'Miljøkraft i Costa Rica', *Teknisk Ukeblad elektroniske magasin* (15), 23 April.

Anonymous (1997b), 'TED Case Studies: Merck-INBio Agreement', URL: http://www.gurukul.ucc.american.edu/TED/MERCK.HTM.

Anonymous (1997c), 'Chiquita-Bananen tragen jetzt ein Umweltzertifikat', *Frankfurter Allgemeine Zeitung*, Frankfurt/M (53), 4 March.

Anonymous (1998), 'Reduction of Methane Emissions from Coffee Mills in Costa Rica', *Joint Implementation Quarterly* 4(4), December, 12f.

Avalos Rodríguez, Angela (1997), 'Pugna por trato con bananeros', *La Nación*, San José (19 March 1997), p.8.

Barres, Herster (1998), 'The Klinki Forestry Project', *International Partnership Report*, 4(1).

Burkard, Christoph (1996), 'Unter einem Sombrero? Massenhafter 'Öko'-Tourismus in Costa Rica', *Blätter des iz3w* (214), June, pp.20–21.

Butterfield, Rebecca P. (1994), 'Forestry in Costa Rica: Status, Research Priorities and the Role of La Selva Biological Station', in Lucinda A. McDade (ed.), *'La Selva: Ecology and Natural History of a Neotropical Rainforest'*, Chicago and London.

Casa Presidencial de Costa Rica (1997): 'Estados Unidos respalda transporte eléctrico en Costa Rica', 15 April, URL: http://www.casapres.go.cr/ visita/electri.htm.

Center for Sustainable Development in the Americas (CSD) (1996), 'Projects in Costa Rica', 13 December, URL: http://www.ji.org/projects/cr.htm.

Center for Sustainable Development in the Americas (CSD) (1997), 'Joint Implementation in Costa Rica', 2 January, URL: http://www.ji.org/jinews/ji_cr.shtm.

Comisión de Servicios Ambientales del Proceso de Concertación (CSA) (1998), 'Consenso para un Futuro Compartido. Informe Final', San José, URL: http://www.nacion.co.cr/concertacion/amb1.html.

Cordero, Carol (1996), 'Empresa privada operará Geotérmico de Miravalles', *La Nación*, internet edn, San José, 3 September.

Dutschke, Michael (1998), ' Financing Sustainable Development. The Case of Costa Rica', HWWA Report No.186, Hamburg.

Escofet, Guillermo (1997), 'Is Deforestation Going from Bad to Worse?', *The Tico Times* (1432), 25 July, p.12.

Escofet, Guillermo (1998), 'Environment Chief Plans Big Changes', *The Tico Times*, online edn, IV(22).

Escofet, Guillermo (1999), 'Worry over Environment Growing', *The Tico Times*, online edn, V(5), 5 Febuary.

Figueres, C., A. Hambleton, L. Lay, K. MacDicken, S. Petricone and J. Swisher (1996), *'Implementing JI/AIJ: A Guide for Establishing National Joint Implementation programs'*, Washington.

Foro Nacional de la Concertación (FNC) (1998), 'Sistema Integral de Retribución por Servicios Ambientales', San José, URL: http://www.nacion.co.cr/concertacion/ambiental.html.

Foundation JIN (1996), 'Certifiable, Tradable Offsets in Costa Rica', *Joint Implementation Quarterly*, (2)2,2 July.

Fuchs, Jürgen (1997), *Costa Rica–Natur in Zentralamerika*, Berlin.

Gesellschaft für Technische Zusammenarbeit (GTZ) (1996), *Measures to Prevent a climate Change*, Eschborn.

Goldberg, Donald M., Carlos Chacón, Rolando Castro and Steve Mack (1998), *Carbon Conservation. Climate Change, Forests and the Clean Development Mechanism*, Washington, DC.

Gorbitz, Adalberto (1997), 'Costa Rica's "Activities Implemented Jointly" Programme', in Kalipada Chatterjee (ed.), *Activities Implemented Jointly to Mitigate Climate Change. Developing Countries' Perspective*, pp.53–8.

Harris, Brian (1995), 'U.S. Approves "Carbon Sequestering" Projects', *Tico Times*, 1(32).

Hein, Wolfgang, Tilman Altenburg and Jürgen Weller (1994), *Autozentrierte agroindustrielle Entwicklung: Eine Strategie zur Überwindung der gegenwärtigen Entwicklungskrise?*, Hamburg.

Heindrichs, Thomas (1997), 'Innovative Finanzierungsinstrumente im Forst- und Naturschutzsektor Costa Ricas', Studie erstellt vom GTZ-Sektorprojekt Unterstützung internationaler tropenwaldrelevanter Programme, Eschborn.

Herold, Anke (1995), *Joint Implementation im Klimaschutz*, Bremen.

Instituto Costarricense de Electricidad (ICE) (1997), 'Proyecto Geotérmico Miravalles', 15 Febuary, URL: http://newton.dgct.ice.go.cr/ice/mirava.htm.

InterAm Database (1995),'Decreto 7508: Reformas de la ley que autoriza la generación eléctrica autónoma o paralela, N 7200', 31 May, URL: http//www.natlaw.com/cr/tropical/eg/dccreg/dccre1.htm.

Inter-American Development Bank (IADB) (1995), 'IDB Approves $18.7 Million for Wind Turbine Power Plant in Costa Rica', 20 December, URL: http://www.iadb.org/prensa/1995/cp29295e.htm.

Intergovernmental Panel on Climate Change (1996), *Climate Change 1995, The Economic and Social Dimensions of Climate Change*, Cambridge.

International Utility Efficiency Partnership, Inc., Edison Electric Institute (IUEP) (1995), 'Klinki Forestry Project', 19 December, URL: http://www.ji.org/usiji/round2/klinki.htm.

Janzen, Daniel H. (1995), 'Biodivesifix - Costa Rican USIJI-approved carbon sequestration JI offsets to support biodiversity development...' 19 December, URL: http://www.chomsky.adelaide.edu.au/Environmental/Brett/janzen3.htm

Jepma, Catrinus J. (1997), 'The Determination of Environmental and Financial Additionality: 11 Case Studies of Uniform Reporting Format (URF) Submissions', unedited.

Lara, Silvia, Tom Barry and Peter Simonson (1995), *Inside Costa Rica*, Albuquerque.

Lay, Loren, Rafael DeLuque and Mahendra Navarange (1996), 'Joint Implementation in Costa Rica', 12 November, URL: http://www.warm.umd.edu/~lorenlay/proj_mis,htm.

LeBlanc, Alice (1997), *An Emerging Host Country Joint Implementation Regime: The Case of Costa Rica*, New York.

Liddell, Jamie and Guillermo Escofet (1997), 'Carbon Bond Scheme Faces Many Questions', *The Tico Times* (1422), San José, 16 May.

Lund, H. Gyde (ed.) (1998), *Definitions of Deforestation, Afforestation and Reforestation*, Thornwood, Ct.

Ministerio de Planificación Nacional y Política Económica (MIDEPLAN) (1997), 'Sistema de Indicadores sobre Desarrollo Sostenible – SIDES', 1 March, URL: http://www.mideplan.go.cr/sides/.

Mulongoy, Kalemani J., Joyotee Smith, Philippe Alirol and André Witthoeft-Muehlmann (1998), 'Are Joint Implementation and the Clean Development Mechanism Opportunities for Forest Sustainable Management Through Carbon Sequestration Projects?', IAE Background Paper 1, Geneva.

Muñoz, Eduardo (1996), 'Una ley forestal para empresarios', *Semanario Universidad*, (4), San José, 12 April, pp.2f.

Muñoz, N., Miguel (1997), 'Gobierno promueve vehículos eléctricos', *La Nación*, San José, 19 July, p.3.

Nilsson, Sten and Wolfgang Schopfhauser (1995), 'The carbon-sequestration potential of a global afforestation program', *Climatic Change* (30), 4/1995, 267–93.

Nordic Council of Ministers (1996), Joint implementation of commitments to mitigate climate change, Copenhagen.

Notimex (1993), En menos de tres decadas podrian agotarse los bosques Costaricenses, *El Financiero*, 21 January.

Oakes, Pamela (1996), 'TED Case Studies: NAFTA and The Environment', 11 January, URL: http.//gurukul.uc.american.edu/ted/COFFEE.HTM.

Oficina Costaricense de Implementación Conjunta (OCIC) (1997), *Costa Rican Certifiable, Tradable Greenhouse Gas Offset*, San José.

Orlebar, E. (1994), 'Call for "carbon bond"', *Financial Times*, 1 June.

Panos Institute (1996), 'Ecotourism – Paradise gained, or lost?', 4 March, URL: http.//www.oneworld.org/panos/panos_eco2.html.

Quesada, Laura (1997), 'Portillos legales facilitan tala masiva', *La República*, San José, 12 July, p.6.

Saborio Valverde, Rodolfo (ed.) (1997), 'Constitución Política de Costa Rica del 7 de noviembre de 1949', 9 January, URL: http://www.nexos.co.cr/cesdepu/nbdp/copol2.htm.

Saito, Junko and Odera Odenyo (1997): 'TED Case Studies: Pesticide Hazard in Costa Rica', 11 January, URL: http://gurukul.ucc.american.edu/ted/COSTPEST.htm.

Santiago, Antonio and Jay Allen Schmidt (1994), 'TED Case Studies: Costa Rica Beef Export', URL: http://gurukul.ucc.american.edu/ted/COSTBEEF.htm.

Scharlowski, Boris (1996), 'Die Jungfrau mit der Banane. Der Chiquita-Konzern versucht ein eigenes Öko-Markenzeichen zu etablieren', *Die Tageszeitung* (4846), Berlin.

Segnini, Gianni (1997), 'Adjudicación del proyecto Miravalles III. Directivos del ICE enfrentados', *La Nación,* internet edn, San José, 2 April.

Société Générale de Surveillance (SGS) (1998), *Certification of 'the Protected Area Project' (PAP) in Costa Rica, commissioned by OCIC (the Costa Rican Office for Joint Implementation), Carbon Offset Verification Report*, Oxford.

Stevens, Mark P. (1996), 'TED Case Studies: Costa Rica Eco-Tourism', 30 April, URL: http://gurukul.ucc.american.edu/ted/COSTPEST.htm.

Subak, Susan (1998), 'Forest Protection and Reforestation through AIJ: Evaluation of the Costa Rica–Norway Project', Working Paper 65, International Academy for the Environment, Geneva.

Sutherland, R. (1996), 'The economics of energy conservation policy', *Energy Policy,* (24), 361–70.

Tattenbach, Franz (1996), 'Carbon Fixation in Costa Rica', *Joint Implementation Quartlerly*, 2(2), Groningen, 1 June, pp.5f.

Tenenbaum, D. J. (1996), 'The Greening of Costa Rica', *Business and Society Review,* (1), Thousand Oaks, California, January.

Trexler and Associates Inc. (TAA) (ed.) (1995): 'The Ecoland Project', 22.8.1995, URL: http://www.teleport.com/~taa/ecoland.htm.

UN Framework Convention on Climate Change Secretariat (UNFCCC) (1997), 'List of AIJ Projects', URL: http://www.unfccc.de/fccc/ccinfo/aijproj.htm., accessed 14 October.

UN Framework Convention on Climate Change Secretariat (UNFCCC) (1998a), 'List of AIJ Projects', URL: http://www.unfccc.de/fccc/ccinfo/aijproj.htm.

UN Framework Convention on Climate Change Secretariat (UNFCCC) (1998b), 'Activities Implemented Jointly – Costa Rican Office on Joint Implementation (OCIC)', URL: http://www.unfccc.de/fccc/ccinfo/aijprog/aij_pcri.htm.

USEPA (1998), AIJ: 3rd report to the UNFCCC secretariat, vol. 2, December, Washington.

US Initiative on Joint Implementation (USIJI) (1994), 'Statement of Intent for Sustainable Development Cooperation and Joint Implementation of Measures to Reduce Emissions of Greenhouse Gases', URL: http://www.ji.org/usiji/la.htm., accessed 9 June, 1995.

US Initiative on Joint Implementation (USIJI) (1996), Activities Implemented Jointly: first report to the secretariat of the United Nations Framework Convention on Climate Change, submitted by the Government of the United States July 1996, Washington.

US Initiative on Joint Implementation (USIJI) (1998), 'USIJI Project Fact Sheet', September, Washington.

Witthoeft-Muehlmann, André (1998), 'Carbon sequestration and sustainable forestry: an overview from ongoing AIJ-forestry projects', IAE Working Paper W75, Geneva.
World Bank (1996), 'Project title: Small and medium scale enterprise program replenishment', 1.10.1996, URL: http://gopher.worldbak.org/html/gef/wprogram/1096/smere2.htm.
World Business Council for Sustainable Development (WBCSD) (1997), 'Proposal Response Form: Klinki Forestry Project', 1 March, URL: http://www.wbcsd.climatechange.com/tasdform/21d6.html.

7. The Impact of Climate Cooperation on Renewable Energy Technologies

Karsten Krause

INTRODUCTION

Access to energy is indispensable for cooking food or boiling water, for heating and cooling. Energy can, indeed, be viewed as a basic human need. Lighting enables an increase in daily productivity, and electricity is vital for most activities in the service sector and industry. A cheap and stable energy supply is thus a prerequisite for social and economic development – in the household as well as at the national level. But the benefits of modern energy are distributed unevenly around the world and have yet to reach approximately one third of the earth's population (UNDP, 1997, p.1). About three-quarters of the world's energy is currently consumed by the industrialized nations, and it is expected that global energy consumption will rise by 30 to 95 per cent in the next 20 years, as a result of population growth and increasing economic activity (Miljø og Energi Ministeriet, 1996, p.18). This increase will exceed the capacity of many energy sources and the capacity of natural sinks for the related emissions.

About 90 per cent of the world's energy is derived from burning fossil fuels (IEA, 1998) and the subsequent emission of energy-related carbon is the central cause of anthropogenic climate change. For this reason, an improvement in the methods of producing and consuming heat and electricity[1] is the core concern of international climate policy. From a scientific point of view, more than a 50 per cent reduction in the global emission of greenhouse gases (GHG) is required to reduce the risk of global warming (IPCC, 1996). International climate policy has thus identified strategies which respect both the increasing demand for energy and the need to reduce emissions significantly. The exploitation of non-emitting energy resources – such as renewables – provides, in this context, an opportunity to merge these contradictory demands.

1. For methodological reasons, the following text focuses mainly on electricity.

Renewable energy technologies (RETs) transform solar,[2] geothermal, and tidal energy into electricity and thus lead to a reduction in the emission of greenhouse gases, and particularly of carbon dioxide. Renewable sources of energy provide the only available option for generating electricity without damaging the global environment. They also avoid the often unacceptable risks which accompany conventional methods of energy production. The most important sources of renewable energy are currently hydro power, wind and biomass, as well as various forms of solar energy. With the exception of large hydro power stations, their exploitation using modern technology is relatively recent. At the moment, renewable energy provides only about 3 per cent of the world's total primary energy supply of 9400 million tons of oil equivalent per annum (International Energy Agency, 1998).

During the 1970s and 1980s, energy security, air pollution, acid rain and nuclear fears were the dominant catalysts for the development of renewable energy sources. In the 1990s, the need to reduce GHG emissions has become increasingly important. With oil prices at their lowest for decades, climate change has become a key issue behind a policy drive which encourages renewable energy (Flavin and Dunn, 1998). A strong incentive for promoting technological development is now inherent in the energy and environmental policies of the industrialized nations. As a result, some technologies are able to compete with conventional sources of energy. Market competition, on the other hand, reflects the interdependence between energy and economic and environmental policies. While the environmental and social costs are basically externalized, different forms of subsidies are available for RETs. Thus RETs are promoted as substitutes for existing facilities in industrialized countries. But, in the developing world, they could also contribute to an environmentally sound and sustainable expansion of the energy system.

The Kyoto Protocol, with its binding commitments for industrialized countries and the introduction of cooperative instruments, could give an enormous boost to the market development of RETs. GHG emission credits for CDM projects are an additional incentive for the application of renewables in developing countries. This may open business opportunities for companies in developing countries and help increase the standard of living in rural areas. Furthermore, markets for RETs in developing countries offer export opportunities to companies from the industrialized countries.

The present chapter analyses the impact of cooperative instruments for RETs in a global context. The starting-point is a general introduction to renewable sources of energy, before the situation of RETs in different countries with their specific driving forces is taken into consideration. The

---

2. Despite photovoltaics to convert solar energy directly into electricity, other energy technologies use sunlight more indirectly. Wind energy or hydro power convert other solar-induced energy flows.

generation of electricity without GHG emission is only one side-effect of RETs. A detailed analysis is provided. The Kyoto Protocol and cooperative instruments have an increasingly important role to play for renewables. Despite the obvious cost-efficiency of trans-boundary offsets of GHG emissions, the long-term effects are more ambivalent. It is thus necessary to analyse, by means of a dynamic evaluation of cooperative instruments, their impact on market penetration, the state of technical progress and the potential to leapfrog market barriers. Finally, both short-term and dynamic effects have to be promoted as developments towards a sustainable energy system.

RENEWABLE ENERGY

General Assumptions

Renewable energy technologies utilize inexhaustible natural resources. They combine environmentally friendly energy production with positive socioeconomic effects: a reduced dependence on imported fuels, new business opportunities for rural areas, and positive employment effects following the production of the equipment in small and medium-sized enterprises. In Denmark, for instance, the manufacturing, maintenance and installation of wind turbines and the corresponding consultancy services currently account for some 10 000 jobs. The export of component supplies and the installation of Danish turbines has created an additional 6000 jobs worldwide (Danish Wind Turbine Manufacturers Association, 1998).

The starting-point for most renewable energy technologies has been the provision of public funds, motivated by the oil crisis of the 1970s. Research and development (R&D) programmes and experimental projects enabled many RETs to enhance their technical performance, but the widespread commercialization of such new technologies has not yet materialized, and this is where the weakness of many RETs currently lies. Low price levels on international energy markets and the externalization of the environmental and social costs of conventional energy are the main handicap for RETs at present.

Figure 7.1 summarizes the expected cost reduction for grid-connected renewable energy technologies by the year 2020. The price array for a particular technology presents the different site conditions for individual projects, such as variations in wind speed or sunlight intensity. The decreasing cost levels in Figure 7.1 show the development of market prices during cost-oriented competition between renewable, fossil and nuclear energy technologies. Despite predictions of further technical progress, only a small share of this potential will be utilized by 2020. But rapid progress in the area of wind power and other technologies has exceeded all expectations and shows how limited such prognoses can be. Wind turbine prices, for example, have fallen by a factor of at least three during the period 1981 to 1991, while energy prices have halved in the last nine or ten years (Bourillon, 1998, p.4).

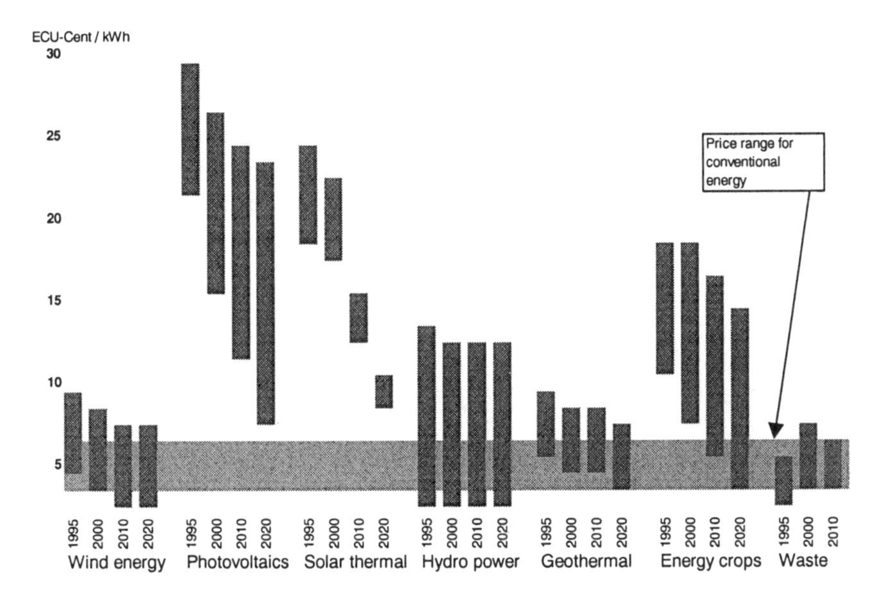

Source: European Commission (1996, p.16).

Figure 7.1 *Price Developments of grid-connected renewable energy technologies*

From an economic point of view, it is also impossible to assign a single value to a unit of generated electricity. Electricity, as a commodity, cannot be stored on a large scale; it can only be generated and consumed on demand. Facilitators thus have to ensure a steady supply which corresponds to variations in daily and seasonal energy requirements. In order to achieve the lowest cost energy supply, it is therefore necessary to provide a number of power plants with differing cost curves, which are able to respond to both base loads and peak demands. With few exceptions, the energy yield of renewables is, however, linked to natural variations. Renewable energy sources are specifically characterized by an uncertain load factor, high initial capital costs and low or non-variable costs. Therefore potentials as well as limitations have to be recognized if one intends to integrate renewables into an existing energy network (see Table 7.1). The value of a kWh, for example, varies significantly with available distribution possibilities and the utilization of economies of scale. Nuclear power stations, for instance, are typical base-load capacities in centralized energy systems. High capital costs allow a 'mass production' of 1000 megawatts and more. Renewable energy is more 'home-made' and benefits from smaller, decentralized systems.

With the exception of large hydro power, the competitiveness of renewables increases as the demand for electricity declines. Many decentralized or stand-alone energy systems already represent a least-cost alternative, even if batteries or other technologies have to secure a stable supply. An example would be the photovoltaic systems on electronic parking meters where, even in the urban areas of industrialized countries, the costs of a grid connection are higher than those of PV electricity, including the battery.

Table 7.1 Energy systems

Centralized electricity generation	A series of large power plants of up to 2000MW, fed by fossil fuel, nuclear or large hydro power and usually sited close to where these sources, or cooling water, are available. The power they generate is transmitted at high voltage to the major demand centres. Distribution systems then diffuse this power at lower voltages for the final consumer. Large amounts of power can be transported across long distances.
Distributed generation	Installation of small generator units in the distribution networks, remote from the large power plant, where they can provide power economically and improve the reliability of the system. Distributed generation is attractive for utilities, and independent power producers compete on deregulated markets. Small units can be added to power systems in increments, avoiding the need for huge one-time investments. Central stations still provide the majority of energy requirements of all consumers on such a system.
Decentralized power systems	Power is provided by a new generation of technologies, including gas turbines, wind turbines, fuel cells and solar generators, which can create electricity locally, even within individual buildings. Thousands of individual generators and consumers are integrated into a single 'smart' system that balances supply and demand by automatically turning individual generation and consumption devices on and off.

	The concept has great potential in sparsely populated areas where few people have access to electricity and where the resources for major power plants are not available.
Stand-alone systems	Small decentralized power systems, often backed up by batteries or diesel generators to maintain a steady supply.

Source: Milborrow (1998); own additions.

Any comparison of costs thus only represents a rough approximation of the technology's value. As many renewables gain less from economies of scale than fossil fuels or nuclear plants, users can deploy them in relatively small sizes. Perhaps more importantly, renewables prove particularly valuable where transmission grids operate at full capacity or where they are situated geographically distant from the actual demand. In short, even at today's prices and with respect to future technological developments, renewables are in a position to provide a valuable service in several markets. Through the exploitation of commercial opportunities, and with substantial political back-up, renewable energies may have the capacity to contribute to about 50 per cent of the global energy supply by 2060 (Deutsche Shell AG, 1998, p.6).

Two separate dynamic currents will shape the present and future application of renewable energy technologies, either through individuals or through the state: renewables as an environmentally friendly source of power, or renewables as a cheap and efficient source of power which is beneficial to development targets. Environment-driven and development-driven demand will be discussed below.

Environment-driven Demand

The control of local air pollution, associated with the burning of fossil fuels, has been one of the primary concerns behind environmental regulation of the energy sector. Minimum standards or emission quotas were established to reduce the impact, but it soon became obvious that end-of-pipe technologies merely shift environmental problems to another ecological dimension. The recognition of the anthropogenic climate change, and the absence of appropriate end-of-pipe technology, demonstrated the need for non-polluting sources of energy. Nuclear power has been promoted as such an energy technology. The risk of nuclear accidents, however, and the existence of radioactive waste has led to public resistance at the same time as national utilities favoured their introduction. The performance of nuclear power plants depends largely on favourable market regulations and subsidies. In the USA, for instance, the deregulation of the energy markets showed the non-performance of these technologies in a competitive environment (Milborrow, 1997).

Environmental and anti-nuclear issues in industrialized countries were the dominant catalysts for the political support for RETs, following the significant decline in energy prices during the early 1980s. In the 1990s, the need to reduce GHG emissions has additionally become an increasingly important impetus for renewable energy (Flavin and Dunn, 1998).

In several industrialized countries parliaments responded to the public outcry and introduced policies to promote the alternative energy sources. Such policies defined minimum quotas for emission-free energy, grants, tax incentives for investments in renewable energy or minimum prices. For instance, in Denmark, Germany and Spain legal pricing regulation (feed-in prices) allows producers of renewable electricity to receive attractive prices for each kilowatt hour fed into the grid of the utility. In addition, strong and consistent tax incentives are bolstering the investment in renewable energy. Wind energy has benefited most from such incentives. In Europe, the annual installed capacity grew from 193MW in 1991 to 1265MW in 1997. By the end of 1997, the installed wind farm capacity in Europe had reached 4425MW, about 75 per cent of the world's installations (Bourillon, 1998, p.11). In response to such policies, some technologies achieved significant cost reductions and became attractive business areas.

The environment driven demand for RETs is concentrated mostly in industrialized countries. Such markets are characterized by four common elements.

1. There is no need for additional electricity-generating capacity. In the 1980s, overestimated demand increases often led to an excess of supply. Also various energy-intensive industries are in decline, while service-based branches are profiting from the modernization process.
2. Monopolistic market structures are increasingly being deregulated, so as to increase the cost-efficiency of the energy sector.
3. Private and governmental institutions in industrialized countries avail themselves of the financial resources to promote renewable energy.
4. There is political interest and international obligation to reduce carbon emissions.

Environment-driven demand for RETs in developing countries is characterized by its focus on local problems. While the global climatic effects of such energy systems are currently neglected, local health problems are recognized as an encumbrance to political priorities regarding economic development. For instance, respiratory diseases, caused by serious air pollution, have become the most common cause of deaths in Chinese urban areas (Preuß, 1997, p.18). The induced loss of human labour and the costs associated with energy-related health problems are increasingly being recognized as a disadvantage in economic development. Therefore RETs are considered an option to balance the expansion of the energy supply with local pollution control.

Development-driven Markets

While energy systems in industrialized countries are slowly changing towards an environmentally friendly structure, the prospects for countries in transition, and particularly in developing countries, are different. At present, roughly two billion people around the globe have no access to electricity at all; other households are nominally served, but the service is so unreliable that they choose to invest in their own sources of power (Kotzloff, 1998). In the absence of reliable grid power, residents become 'self-providers': they use diesel generators, kerosene lamps, lead acid batteries charged by diesel generators, candles and diesel pumps. Self-generation constitutes an average of 13 per cent of the total power generation in 75 developing countries with available data, and represents over 25 per cent in 12 of those countries, mainly in Africa (ibid.). Many of these sources emit pollutants with adverse environmental and health effects.

The existing centralized energy systems in many developing countries are incapable of providing sufficient electricity. An increase in population and economic growth will widen the gap between the phases of power delivery if supply capacities are not expanded. In China, for instance, the Ministry of Energy has estimated that demand is outgrowing availability by 15–20 per cent. By the year 2015, the gap will have increased by 150 per cent – if the current political reforms continue (Kusch and Blank, 1997, p.3). Typical of the emerging markets in Asia, the Chinese energy system is based on coal-fired thermal plants with very low efficiency and no adequate scrubber systems. Outdated technologies, unreliable and low fuel quality and considerable management problems are increasing emission intensity significantly. Owing to the high growth rates of electricity demand – double-digit figures in some countries – new investments in the power sector in developing countries over the next 20 years will amount to the cumulative total that has been invested to date (Kotzloff, 1998).

The current situation is ideally suited to the introduction of RETs. Expected reforms and the necessary expansion of the power sector in developing countries will open huge markets for renewable energy technologies. Renewable resources are indigenous, do not require fuel purchases and can be used locally for power generation. So RETs are particularly advantageous for off-grid applications. Nevertheless, developing countries tend to emulate electricity sector models that were pioneered in industrialized countries. Even after their independence from colonial rule, they sought the implementation of a centralized energy system like those in Europe, but, as result of a lack of indigenous capital and technological experience, expansion plans have not been carried out successfully.

Until recently, the rare deployment of renewable energy technologies in developing countries occurred primarily through direct government investments or international donor programmes in which renewables did not compete in open markets. Where renewable pilot projects have been implemented by publicly owned and managed utilities, the decision was often made in order to meet planning, political or technological objectives.

By following the model of the industrialized countries, and the advice of international finance institutions, the public structure of energy utilities has now come to undergo a transformation towards deregulation and decentralization. New, competitive market structures will open new opportunities for RET-based projects. The current situation is ideally suited to the introduction of RETs. The expected reforms and the necessary expansion of the power sector in developing countries will open huge markets for renewables. In some cases, renewables will even have a significant competitive advantage, particularly in rural areas.

But this window of opportunity will eventually pass. If the developing nations adopt rules that tie them to conventional technologies, they will lose a unique possibility to develop an environmentally and economically efficient power sector. Once installed, a centralized energy system based on conventional technologies, will limit the market potential for RETs and increase the perspectives for centralized fossil fuel plants. The future market penetration of renewables in developing countries depends on which model developing countries emphasize for their power sector expansion and, of course, on the influence of industrialized countries, for example the availability of assistance for national energy policies or the demonstration of a development-oriented use of RETs in their own energy sectors.

COOPERATIVE INSTRUMENTS

Renewable Energy and Global Warming

There are four different options for the reduction of GHG emissions.

1. The transition to fuels with lower specific emission intensities, such as the replacement of lignite by natural gas or by renewables.
2. The implementation of more efficient technologies in order to maximize the energy output per GHG emission unit. Developing countries, in particular, use outdated, inefficient technologies in their energy systems.
3. A reduction of energy-intensive activities, initiated by price incentives, public awareness campaigns and so on, and a strategic acceleration of the modernization process. An example of such a strategy is the internalization of external costs by carbon taxes or tax incentives for investments in environmentally friendly technologies.
4. The fourth option focuses on limiting the GHG concentration in the atmosphere, by binding it in biomass. While carbon emissions cannot be prevented efficiently by filter technologies, changes in land use and afforestation programmes can sequestrate carbon emissions. The IPCC has pointed to an important role for renewable energy in eventually meeting the goal of virtually replacing fossil fuels, noting that, 'in the longer term, renewable energy sources could meet a major part of the world's demand for energy ' (International Panel on Climate Change, 1996).

Despite the importance of inexhaustible energy resources for future generations, climate policies have to balance this potential with existing budget restraints. In order to achieve GHG emission reduction targets over a given period of time, cost-efficiency takes on an increasing significance. Within any one country, the government has the opportunity to initiate and control the four different emission control options. The choice will depend on political considerations and on the marginal costs associated with offsets. As a consequence, the cost per unit of carbon emission reduction is important for a climate policy-motivated application of renewable energy technologies. Figure 7.2 shows the costs per ton of CO_2 emission reduction in Germany.

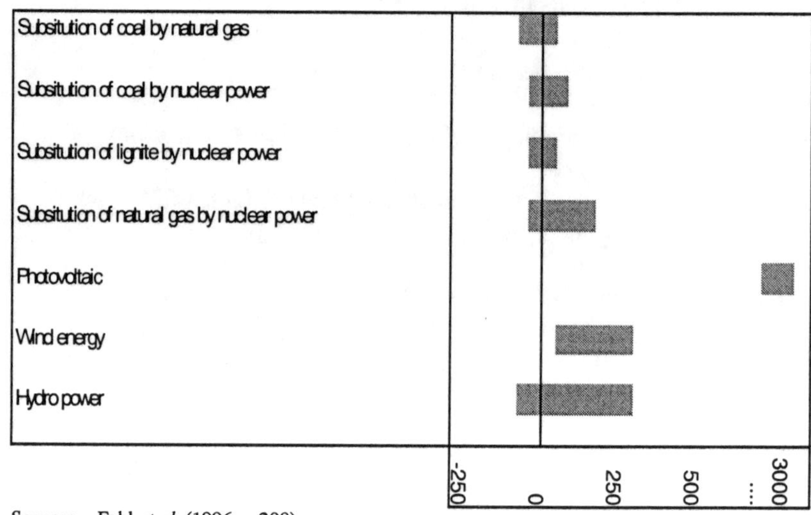

Source: Fahl *et al.* (1996, p.208).

Figure 7.2 Comparison of costs for CO_2 emission reductions of different energy technologies (based on 1990 price levels)[3]

Owing to rapid progress in the development of energy technologies, and renewables in particular, their offset costs have fallen since the early 1990s. Nevertheless, low prices for fossil fuels like natural gas limit the role of renewables as a competitive option for a cost-oriented climate policy. In addition, energy-efficient technologies and changes in land use enable the reduction of emissions at low cost, often lower than renewable energy technologies.

There are significant variations in the cost of energy production, depending largely on the location and time of electricity supply. Among such variations, however, some renewable energy technologies find attractive market niches from which they can compete with conventional fuels. The offset of GHG emissions is indifferent to such price variations. On a level playing-field, only attractive costs per emission reduction unit will allow renewables to utilize cooperative instruments to accelerate their market penetration. The future performance of renewables will therefore be sensitive to the formulation of regulations which shape the global market for emission reduction units.

3. Negative cost indicate the viability of an energy technology. If, for instance, a coal-fired plant is replaced by a natural gas station, the costs per kilowatt hour will decrease and carbon emission reductions are just a side effect.

Flexible Instruments and Renewable Energy

One of the key subjects of the discussion in Kyoto and in Buenos Aires was the introduction of cooperative instruments. Provisions for international cooperation in meeting emission reduction commitments are a novel and crucial feature of the Kyoto Protocol. As discussed in previous chapters, the Kyoto Protocol institutionalizes four mechanisms of cooperative implementation: emission trading, joint commitments, the clean development mechanism (CDM) and joint implementation (JI). Table 7.2 summarizes the different options for countries with binding commitments to achieve compliance with the GHG emission reductions as agreed in Kyoto.

The rules for the different mechanisms are currently vague and need to be more precise in order to ensure that economic efficiency and environmental sustainability are promoted at the same time. The facilitation of a transboundary sharing of the costs of climate policy and the mobilization of a transfer of private capital and technology towards developing countries and the economies in transition will have an impact on the future of RETs.

For renewable energy technologies, the Kyoto Protocol, with its binding commitments for industrialized countries and the introduction of cooperative instruments, promises both new markets and competition for cheap emission reduction units. There are many possible combinations of energy saving and fuel substitution that would meet the Kyoto commitments, but all involve large deviations from past energy trends. Major new policies and measures need to be put in place urgently by governments to meet the Kyoto objectives. In these policies, RETs have to compete against other energy technologies, such as natural gas or nuclear fission, against energy efficiency measure or against afforestation projects. The role of RETs depends largely on the specification of the rules for the individual cooperative instruments. The following rules are of particular interest.

Table 7.2 Possible methods of implementing the commitments made by Annex B countries under the Kyoto Protocol

		Principle	Participants
National enforcement		Transmission of the Kyoto commitments into the economies of Annex B countries in compliance with existing national regulations.	Individuals, private and public sector entities. Example: carbon taxes in Germany.
Flexible Instruments	Joint commitments (Article 4)	A group of Annex B countries fulfil their Kyoto commitments jointly, if their total combined aggregate emissions do not exceed their assigned amounts. This must be declared before ratification. The agreement cannot be revised during the commitment period.	Governments of Annex B countries. Example: EU bubble policy.
Flexible Instruments	Emission trading (Article 17)	Provides the possibility of trading emission reduction and limitation obligations. Trade-ins are permitted only to fulfil the countries' commitments and must be supplementary to domestic actions.	Governments of Annex B countries. Example: sale of Russian 'hot air'.
Flexible Instruments	Joint implementation (Article 6)	Annex B countries are allowed to transfer or acquire emission reduction units resulting from emission reduction projects. Countries may also authorize legal entities to participate, under the country's responsibility, in actions leading to the generation, transfer, or acquisition of emission reduction units.	Private and public sector entities can participate in JI activities. Example: replacement of coal-fired power plants in Russia.
Flexible Instruments	Clean development mechanism (Article 12)	CDM allows non-Annex B countries to sell certified emission reductions obtained through project activities. A share of the proceeds will be used to help particularly vulnerable developing countries meet the costs of adapting to climate change.	Private and public sector entities can participate in CDM activities. Example: replacement of coal-fired power plants in China.

Supplementarity to activities in industrialized countries

Any voluntary participation by governments and/or individual emitters from industrialized countries is motivated by the expected cost savings from the trans-boundary fulfilment of emission reduction commitments. This contains the risk of a reduced effort to use RETs in the investor country. Any emission reduction that is achievable abroad reduces the need for domestic activities. To prevent negative implications for renewables, caps for emission trading between industrialized countries and an exclusion of 'hot air' and 'tropical air' are essential. Without clear limitation of the trans-boundary fulfilment of emission reduction commitments, low price levels may have negative implications for RETs in developing and industrialized countries.

Offset additionality

Exchanged emission reductions must be additional to what would have been done otherwise. The decision as to whether created entitlements are 'additional' thus requires the existence of a baseline against which the reduction can be measured. Within the definition of a baseline for renewable energy projects, the mix of GHG emission reductions, the value of generated energy, diminished impacts of other pollutants and other effects have to be taken into consideration. When GHG emissions are reduced below this baseline, the exceeding reduction can be certified as additional. Owing to individual differences between projects, the baseline has to be defined, at least initially, on a case-by-case basis rather than at the national level. The definition of additionality is particularly complex in the application of renewable energy technologies. As a result of the interaction of different income sources and various positive effects, it is a complicated matter to analyse the motivational structure behind each project.

Transparency and reliability

In order to ensure the supplementarity and additionality of renewable energy applications, an institutional framework for cooperative instruments has to be established. It has to prevent loopholes and, at the same time, it should minimize transaction costs. Overcontrolling and bureaucratic procedures for any trans-boundary interaction will discourage potential participants, in particular from the private sector. Unreliable crediting could easily lead to creative bookkeeping, instead of inducing global GHG emission reductions.

While the role of RETs under the Kyoto Protocol will be influenced by the outcome of current negotiation processes, the main effects of the cooperative mechanisms depend on their final design.

Opportunities and Risks

The negotiations at CoP4 in Buenos Aires have not resulted in any fundamental decision on cooperative instruments. Nevertheless, trans-boundary emission offsets are seen as a promising opportunity for private sector initiatives in the energy sector. While some environmental activists fear that cooperative instruments shift the focus on to trading rather than projects that aim at actual and definite reductions in greenhouse gas production, others regard the Kyoto Protocol as a promising market opportunity for renewable energy technologies. Governments in industrialized countries are counting on renewables to fulfil their Kyoto commitments. For instance, the European Union issued a strategy to double the share of renewables in the energy system, with the target of a one-quarter contribution to their commitment (European Commission, 1997).

Despite their role in national strategies to fulfil the Kyoto obligations within industrialized countries, maybe the most promising market opportunity is the project-based implementation of RETs in developing countries. The clean development mechanism (CDM) is potentially one of the Kyoto Protocol's most crucial developments for renewables (Flavin and Dunn, 1998). Unlike other cooperative instruments, the CDM involves private and/or public entities in activities in a decentralized and expanding energy system. CDM is a multilateral financial mechanism, intending to yield project activities which result in emission reductions and a stimulation of economic development in developing countries. Despite uncertainties about the function of the CDM as a quasi-public mutual fund or as a purely private trading system, it has the potential to initiate new business activities. The CDM seeks to establish an international assistance fund for those developing countries that are most vulnerable to climate change and procedures for project assessment and approval that would further the interests of all participants.

Under the CDM, a project developer might build a wind farm in China, generating emission credits that could be sold to a company in Europe, in order to reduce the tax level for its domestic carbon emissions. This new income perspective is an additional incentive for the deployment of renewable energy technologies on the development-oriented energy market. Compared with the renewable energy projects of national or multinational development assistance authorities, the CDM provides market-based incentives for investment in developing countries. Emitters from industrialized countries, or project developers, possibly private companies or environmental non-governmental organizations, may use the CDM. They could either reduce their own emission reduction obligations or sell emission credits on international markets.

Figure 7.3 gives a schematic overview of the integration of the CDM into energy projects in developing countries. In addition to the sale of electricity and subsidies by international financing institutions or bilateral development aid agencies, the CDM is a third potential source of project income.

The CDM brings additional value to RET projects. Its purpose is to generate lower-cost emission reductions for industrialized countries that are currently obliged to reduce emissions. Furthermore, sustainable development benefits for host developing countries, including economic and environmental benefits and the transfer of lower-emissions technology, are achievable. Despite short-term benefits in the fulfilment of GHG emission commitments, the most essential aspect for renewable energy technologies is the dynamic effects.

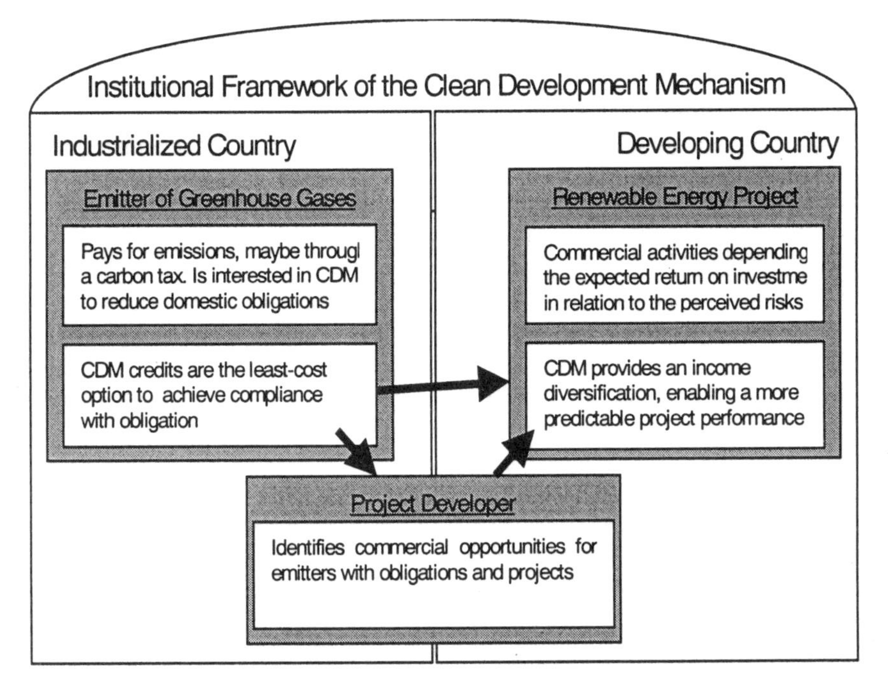

Figure 7.3 Structure of a CDM renewable energy project

DYNAMIC EFFICIENCY

Stimulating Market Penetration

The exploitation of the technical potential[4] for each renewable energy source depends on the payback time of initial project costs. The relationship of capital costs and competition between renewable and non-renewable energy technologies can be simulated in a market potential curve. The range of the S-curve provides the link between different incentives for the initiation of a project development. The longer the payback period, the smaller the market potential, and vice versa. Different shapes of the S-curve indicate different rates of market penetration. A steeper S-curve implies a more rapid rate of penetration, so that the full market potential is reached more rapidly than is the case with a shallower curve.

The shape of the market potential curve (Figure 7.4) is determined by market incentives and the political support for the development of RET-based projects. Limitations derive from the number of available project sites, market barriers and the current stage of technological development. If a market potential exists, its identification by market actors and the initiation of project developments delay the utilization. The market penetration curve indicates the ability of existing market structures to respond to new business opportunities.

For the development of RET-based projects in developing countries, the introduction of CDM provides the opportunity to reduce the payback period and therefore to expand the market potential. The public debate on CDM and the other flexible instruments will increase awareness of these new business opportunities. Increasing market size will attract new actors and help established market actors to improve their services.

An important limitation on the current market penetration of renewable energy projects in developing countries is the concentration of total project costs before the start of operation, and relatively long payback periods. Many hydro power projects have an operating time of 50 years and normal payback periods of 30 years (Feuerstein, 1998). While other technologies, like wind power, are able to amortize in less than 10 years, this makes them vulnerable to a changing project environment. There are three different factors that are of particular importance to the market penetration of RETs. First, for many technologies rapid technical progress makes it impossible to base calculation models on empirical data. Second, the value of electricity depends on the structure of the energy market as well as on international fuel prices and

4. The technical potential is defined as the upper limit for the technical accessible energy from one source, regardless of costs. The market potential recognizes the relationship between a project's payback time and the exploitation of the technical potential.

national energy policy, which might, for instance, favour the expansion of the centralized energy system. Third, the macroeconomic instability of many developing countries might endanger the project's profitability through non-transparent tax and tariff regulations (ibid.). Currency devaluation in particular, such as happened in Asia recently, might have a significant impact on RET-based projects. Currently, the largest share of the required equipment has to be imported from industrialized countries. With respect to refinancing the costs during running time, a declining exchange rate in developing countries prolongs the amortization period.

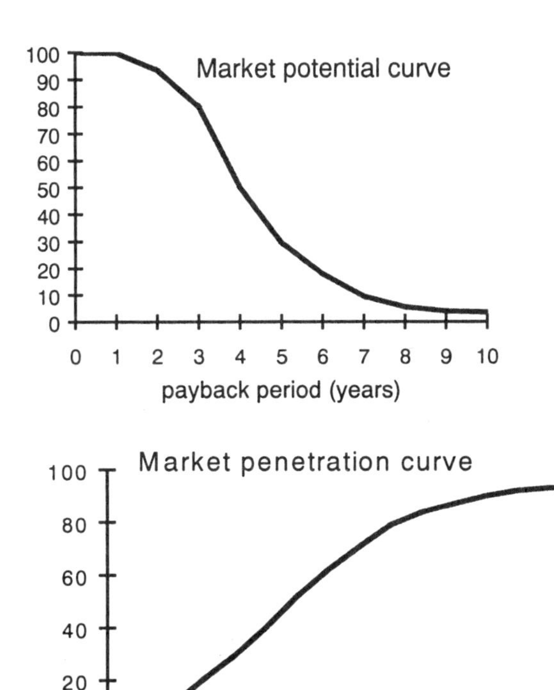

Source: European Commission (1996, p.56).

Figure 7.4 Market potential and market penetration curves

The CDM also has the potential to reduce the impact of the political environment on projects in developing countries: the additional income opportunity from the sale of emissions credits to industrialized countries reduces the revenue from electricity sales to only one possible source of income. Basically, the CDM-induced stream of foreign currency revenue significantly limits the impact of political instabilities on the project. The repayment of the initial project investments requires the availability of foreign currency. Depending on the individual financing model, the sale of CDM credits can be used to repay the initial costs more directly. This opportunity may lead to a different risk analysis by potential project investors. Investors could then be found internationally, but also locally – depending on the individual financing model.

One example of the successful exploitation of an available market potential is the activities of the Solar Electric Light Company (SELCO) from the United States. SELCO promotes solar home systems, consisting of a small solar panel, a battery and energy-efficient appliances (lights, radio) in developing countries. Most of the 400 million households which currently have no access to electricity will never be connected to any kind of centralized grid system, or at least not in the next decade. In the absence of a centralized energy supply, consumption via an electricity grid is only possible if access to energy using alternative methods is organized and financed. Often, the monthly expenditure for candles, batteries or kerosene accumulates to more than US$10 (SELCO, 1998, p.3). In addition, there are an estimated 50 million rural shops, clinics, schools, community centres, banks, cooperatives and other small enterprises whose productivity could be improved by stand-alone systems (ibid., p.3). Figure 7.5 shows how the chosen form of financing influences the market penetration of solar home systems in developing countries.

There are about 200 million households that currently pay for energy services and which could afford the costs of a solar home system. The need to purchase equipment before the start of the operation is the impregnable market barrier for 95 per cent of the potential customers. If credit financing is an available option for purchasing the system, the market potential can grow by a factor of ten. Still, 45 per cent of these households would need additional subsidies to make solar energy attractive to them.

SELCO sees the CDM as an opportunity to link its activities in the marketing and installation of solar home systems in rural areas of developing countries.[5] In conjunction with a number of partners, it started a pilot project carried out under the United States Initiative for Joint Implementation. Over 812 000 solar home systems are to be installed over a 10-year period in Sri Lanka. These systems will replace the use of kerosene lamps for lighting and the use of diesel electric charging of lead-acid batteries for powering small

5. SELCO currently operates in India, China and Sri Lanka. It is expanding rapidly into other countries, for example recently Zimbabwe.

home appliances. It is expected that each system will generate GHG benefits over 20 years. A successful project will lead to cumulative savings of over 6 million tons of CO_2 (Lile *et al.*, 1998, p.23).

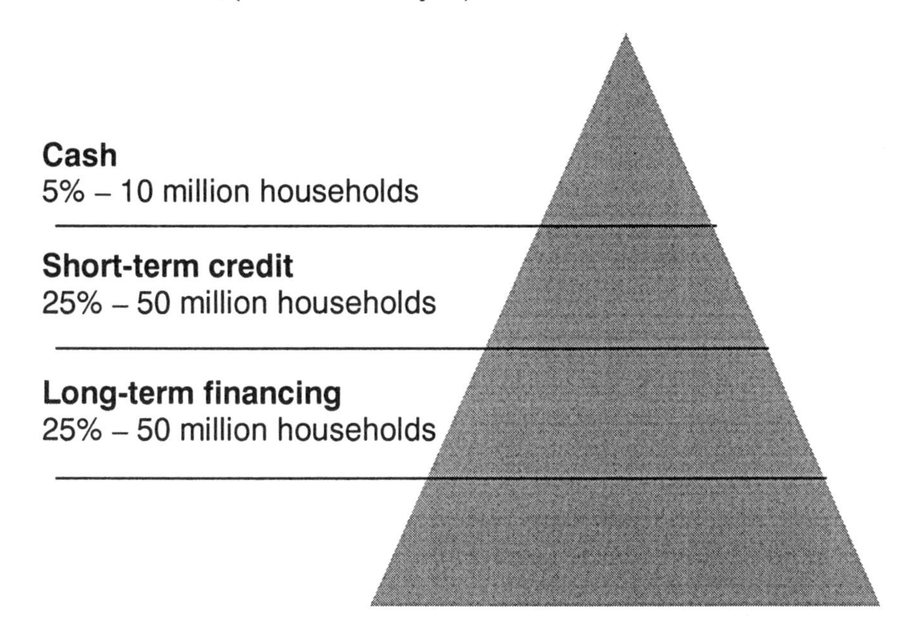

Source: SELCO (1998, p.3).

Figure 7.5 Global market potential for solar home systems

Within this pilot project, it is difficult to define the additional incentive offered by CDM credits. The project involves significant risks for any commercial investor. The distribution of the solar home systems to individual households adds a substantial amount of transaction costs to the expenditure for the technical equipment. Identifying 812 000 households, negotiating the contracts and ensuring the repayment over 20 years requires additional resources.

For projects that set out to promote small-scale technologies, transaction costs are a significant limitation on market penetration in developing countries. Projects like SELCO's Rural Electrification Programme have to develop long-term financing models, leasing arrangements or fee-for-service concepts to refinance the initial investment costs of US$300–500. Additionally, many people in developing countries have little or no experience of RETs. In order to explain the advantages of the new technology and to implement it in existing social structures, awareness-raising campaigns are necessary. Investments in such measures are viable if a minimum market potential is available.

The availability of CDM credits can make RETs such as solar home systems affordable to a wider range of consumers. This would help to distribute the project costs better and to reduce the risk of non-performance of the project marketing. For many projects, the CDM-induced enlargement is a prerequisite for reaching the critical mass that allows commercially viable operations.

Accelerating Technical Progress

The long-term aim of international climate policy is the stabilization of global GHG concentrations – at minimum costs and maximized synergy effects with other policy areas. Cooperative instruments can provide higher efficiency in the implementation of short-term national commitments. Nevertheless, short-term efficiency has to be balanced with the long-term process of developing an energy system that guarantees future generations security of supply and prevents ecological risks (Michaelowa and Schmidt, 1997). These requirements highlight the importance of technical progress in existing renewable energy technologies and the development of new inexhaustible energy sources. With the Kyoto Protocol the positive effect of renewables on global warming gained more prominence and gave an impulse to their further development. The question that currently remains open concerns the direction of this impulse: will it encourage the development of those technologies with the best future potential or will it only encourage those which are already at an almost commercial stage of development?

The expectations placed in market dynamics are linked to the assumption that the increased application of a technology will lead to significant cost reductions and initiate technical advance. The experience curve (see Figure 7.6) relates to the various stages of the product life cycle and demonstrates the potential for achieving dramatic cost reductions from experience in producing a new product.

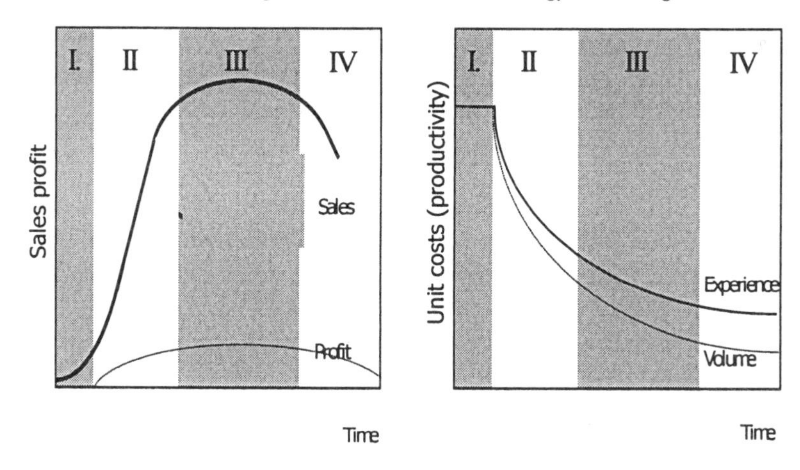

Note: I Introduction, II Growth, III Maturity, IV Decline.

Source: Tatsis (1996, p.11).

Figure 7.6 *Product life cycle: experience and volume curves*

For each doubling of cumulative output (experience) of a new product, there is typically a certain percentage reduction in unit costs. Cost reduction rates vary between products. It is reasonable to expect, in the case of RETs, that final unit cost reductions of between 10 per cent and 15 per cent are achievable (Tatsis, 1996, p.3). As the volume curve follows a similar trend to the experience curve, similar gains might be expected from achieving the benefits of economies of scale (ibid.).

At the moment, most RETs are in the early stages of their product life cycle. As mentioned above, large-scale hydro power is an exception, and in recent years the wind power markets experienced significant growth. The resulting cost reductions led wind farms in good wind conditions and with a capacity of more then 10MW to become the cheapest energy source available (Milborrow, 1997, p.21). Nevertheless, long depreciation periods of the capital stock and different market barriers (see the following section) reduce their implementation rate and consequently reduce technological development and the achievement of economies of scale. The Kyoto Protocol and CDM have, theoretically, the potential to initiate a market pull for RETs: the perspective of emission credits can induce additional demand and help RETs to move into a phase of development where they can grow and expand their market share. The permanent character of the CDM income stream allows market actors a more reliable prognosis of future market development. Today, the availability of grants or soft loans is the bottleneck for most project developments. Consequently, the technical progress of individual RETs depends on political decisions and temporary assistance programmes.

The application of individual RETs under CDM will depend on prices on the international CDM credit markets. If too many cheap energy efficiency or forestry projects offer credits, the low price level may crowd most renewables out of the market. In any case, the market-based approach of CDM will allocate capital to the most efficient project options. This would bring about a differentiation between RETs which will gain from CDM and those with a neutral position. Because of existing transaction costs, large-scale applications of RETs will benefit most from an undifferentiated crediting of GHG emission reductions. Technical progress will induce an increasing bias of capital transfer to the most cost-efficient technologies. This will bring imbalances into the CDM mechanism. Therefore, a sliding reduction of crediting helps to link short-term efficiency gains to the future potential of a technology (Michaelowa and Schmidt, 1997). If individual RETs or specific project types come into a situation where their implementation no longer fulfils the requirement of additionality, the crediting ratio for these projects should be reduced according to their performance improvements and research results. Without control of the technology choices within CDM projects, imbalances may occur and limit the dynamic efficiency of CDM: in order to realize short-run efficiency gains, only the most cost-efficient technologies will be applied. While these technologies achieve significant technical progress, other options fall behind. Without the interaction of different technological options, strategic expansion will depend on available project sites and will make the contribution of renewables dependent on one energy source. Instead of balancing natural load variations between renewables, a back-up by fossil fuels will be necessary. Consequently, a carefully designed framework for CDM has to set strategic incentives for inducing and accelerating innovation for a wide range of renewable energy technologies.

Leapfrogging Market Barriers

Market barriers limit the penetration of a theoretically available market potential by RET-based projects and their achievement of technical progress. Barriers can occur in five interdependent areas: technical, economic, environmental, social and institutional. Technical restrictions depend on the individual technology. They can reduce the security of supply, or the performance on a site. Economic barriers are linked to critical mass factors due to the lack of economies of scale. An important economic barrier is the lack of access to low-interest loans and equity investments. While renewables reduce emissions of GHG, they can affect the local environment in different ways, such as the impact of a hydro power station on the ecosystem of a river or the visual impact of wind turbines. These trade-offs between different environmental impacts and limited public awareness about global warming can affect negatively the market penetration of RETs. But the most important barriers are institutional ones. Often they have developed for historical reasons, such as the protection of the utilities' monopoly. For technologies like RETs and other decentralized energy technologies, the legal framework and market structures are inadequate. The embedding of each project into the existing energy system of the host country may interfere with the project's commercial aims. One of the crucial points is the political determination of energy prices. Owing to the importance of energy prices for regional development and economic growth, different forms of subsidies limit the transparency of price structures. Consumers get no price signals about the costs related to the production of their energy. This lack of information can influence the willingness to pay for electricity from renewable sources. Basically, these barriers occur in both developing and industrialized countries, but with a different emphasis in each case.

The introduction of cooperative instruments and CDM in particular will help tackle some of these barriers, while new ones might occur. The expansion of the number of RET-based projects in developing countries can help with finding new forms of production, project development and financing. Increasing penetration may show limitations of a technology that had not previously been apparent. A rising awareness of the visual impact of wind farms and the need for visual impact studies in countries such as Ireland are an example. In developing countries, institutional deficits and a focus on market potential can easily threaten the residents in the direct neighbourhood of a project site. To avoid conflicts and to maximize local benefits, the social impact will have to be included in project assessments before CDM crediting takes place.

The most important effect of CDM on the energy system in developing countries is the introduction of an environmentally and economically efficient power sector, based on RETs or at least with a significant contribution by them. Without the sunk costs in conventional technologies and with a transfer of capital, technologies and know-how from the industrialized countries, RETs have the opportunity to leapfrog many barriers which they currently face in the industrialized countries.

Additionally, CDM could help to leapfrog market barriers in industrialized countries: through the application of RETs in developing countries, RETs in industrialized countries will benefit as well. This assumption presumes world wide homogeneous technical demands on the implementation of technologies. In fact, fundamental differences between the project environment in industrialized and developing countries can interact with the requirements for an intermediate technology and an adequate project structure.

Up to now, the basis for the development of RETs was the interaction of commercial interests with research institutes, public support and governmental programmes in industrialized countries. Most of the induced technical progress was suited to the conditions on environment-driven markets. Such factors still are the pacemaker for innovations in this area. Within these markets, the deregulation of the energy market and the Kyoto Protocol bring about new opportunities for, and threats to, individual project developments in the existing centralized energy system and the transformation of the market structure towards decentralization.

A significant share of the innovations in this market require a specific infrastructure for their installation and maintenance. For instance, the need for site access for heavy-load vehicles or the recurrent inspections of the operating project can add costs to an individual project in developing countries. These can often be prohibitive for projects in remote areas. Intermediate technologies and project developments have to internalize the scarcity of infrastructural elements that are a matter of course in industrialized countries. Therefore, theoretically possible project sites for small-scale systems with 'plug-and-play' technologies will find the largest market potential. The demand for RETs such as solar home systems in industrialized countries is limited to niche markets.

This need for intermediate technologies in combination with the scarcity of resources can lead to the interpretation of developing countries as inferior markets by equipment manufacturers, trading companies or project developers. On the one hand, it could bring about a transfer of suitable and affordable technologies. On the other hand, this could induce the export of questionable technologies. Technical progress rapidly devalues existing technologies in industrialized countries, limiting manufacturers in the maximization of the profits from established products. The limitation of sites for new projects makes an upgrading of existing equipment attractive to project operators. Both options make it attractive to use developing countries as a market for these old, outdated technologies. CDM could help to install such a transfer, owing to the international linkages associated with many projects.

The markets for RETs in developing countries could also be used to test new, unproven technologies, but this is resisted by the recipients. The promising market and the limited RET experience in developing countries could easily lead to CDM being used as a funding source for a demonstration project the benefit of which to the host country was questionable. CDM-induced RET exports might transfer only current technologies that will rapidly become obsolete or might take the form of a demonstration operation not based on adequate practical experience. It is necessary to control the standard of the technology applied and the beneficiaries of the induced learning effects. In the long run, it is necessary to initiate technologies and project structures that are adjusted to conditions in developing countries. This standard does not need to be identical to that of the market structures in the industrialized countries. To induce the necessary market processes, the export of technologies can be just the first step. The transfer of knowledge and local capacity building have to initiate the development of an independent RETs industry in developing countries.

The best possible support for any renewable energy would be the removal of all subsidies, including hidden subsidies that allow the externalization of the social and environmental costs of energy production. Without such imbalances, a level playing-field would allow the commercialization of most RETs (Tatsis, 1996, p.30). The second-best option is the introduction of subsidies for renewable energy, to limit the remaining effects of existing market barriers. CDM has the potential to reduce the imbalances and can act as a strategic promotion for RETs through the creation of adequate market structures for their commercialization.

TOWARDS A SUSTAINABLE ENERGY SYSTEM?

Sustainable development, as the paradigm for environmental policies, gained prominence at the Earth Summit in Rio de Janeiro in 1992. It is defined as a balance between the economic, environmental and social needs of the present with those of all future generations.

A sustainable energy system enables access to sufficient and affordable energy services and recognizes, at the same time, environmental restrictions. Sufficiency and efficiency are important criteria for both the demand side and the supply side of the energy sector: minimization of primary energy input and supply using the best available technologies (Meyer-Abich, 1995, p.105). While renewables have the ability to provide energy services without the many negative side-effects of conventional technologies, their introduction is not necessarily a 'no-regrets' option. The negative consequences strongly depend on the individual technology. Wind power, for instance, causes a visual impact dependent on the height of the wind turbines. The rotation of their blades may cause sound emissions and disturb the bird life in their direct neighbourhood. Another negative point might be the energetic and environmental amortization[6] of some RETs, in particular photovoltaics. The

6. The production of the technical equipment requires an input of energy and different materials. This causes an impact on the environment. The energetic and environmental

environmental impact of the energetic amortization and the technical components of some outdated PV-products exceeds all the benefits which they generate during their operational time. Another critical point is the price of RET-generated electricity. The costs of using renewable sources to achieve political targets, such as the Kyoto commitments or employment creation, can be relatively high. Compared to cheaper alternatives, and owing to the uncertain long-term effects of RETs, the share of a limited budget spent on renewables can cause the loss of other opportunities and interfere with other political priorities, such as employment policies. Nevertheless, the introduction of CDM and its combination with RETs has the potential to reduce the tradeoff between different areas (see Figure 7.7).

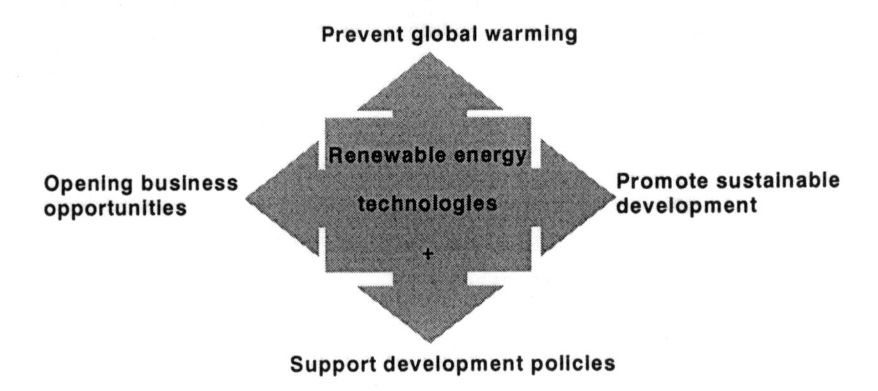

Figure 7.7 Bridging trade-offs through the combination of RETs and CDM

With every kilowatt hour of electricity produced, RETs can replace emission-intensive energy sources. Placing a monetary value on the side-product 'carbon-replacement' of RETs can help to overcome traditional trade-offs: the prevention of global warming becomes a source of income and attracts business interests. Countries can expand their energy production and at the same time promote an environmentally friendly and socially acceptable growth pattern.

Despite the potential to link different interests, the CDM-induced market potential and market penetration can only be one element in the prevention of global warming and the introduction of a sustainable energy system. While the price of GHG emission reduction units depends on the institutional framework of cooperative instruments under the Kyoto Protocol, the

amortization expresses the relation of the initial impact and the positive effect during the operation of the RET.

individual market potential also depends on the energy policy of the host country.

From the analysis of the impact of climate cooperation on different RETs, it is obvious that two factors are crucial for the implementation of a sustainable energy system: (a) the balancing of the commercial incentives of private sector actors with the CDM requirement of additional projects that would have happened anyway; and (b) the political recognition of energy as a basic need, including the individual right to energy access that is required for clean water and food. This right includes the physical and economic access to energy.

The main reason for global warming is the energy intensity of consumption and production in the industrialized countries. A strategic transformation towards sustainable energy systems cannot take place without the implementation of efficiency and sufficiency measures in the industrialized countries. Cooperative instruments may help to limit global warming, but they also reduce the need for domestic strategies. Therefore any RET-based CDM project can never be more than a second-best option, inferior to domestic policies. Among the different technical opportunities for this second-best choice, RETs are doubtless the first choice for meeting the Kyoto objectives and promoting sustainable development.

REFERENCES

Bourillon, Christophe (1998), *Wind Energy – Clean Power for Generations*, London.

Danish Wind Turbine Manufacturers Association (1998), home page www.windpower.dk.

Deutsche Shell AG (1998), 'Globales Marktpotential für erneuerbare Energien', Hamburg.

European Commission (1996), *TERES II – The European Renewable Energy Study 1995–2020*, Brussels.

European Commission (1997), 'Energy for the Future: Renewable Sources of Energy', White Paper for a Community Strategy and Action Plan, Com(97)599 final (26/11/1997), Brussels.

Fahl, Ulrich, Egbert Läge, Peter Schaumann and Alfred Voß, (1996), 'Wirtschaftsverträglicher Klimaschutz für den Standort Deutschland', *Energiewirtschaftliche Tagesfragen*, (4), 208–12.

Feuerstein, Horst (1998), 'Uncertainties of Cash Flow for Investments in Developing Countries', Presentation on the 2nd International EUROSOLAR Conference 'Financing Renewable Energies', 16–18 November, Bonn.

Flavin, Christopher and Seth Dunn (1998), 'Climate of opportunity – renewable energy after Kyoto', www.ewpp.org/articles/issuebr11/.

International Energy Agency (1998), 'Energy Statistics', http//www.iea.org/stats/files/keystats.htm.

International Panel on Climate Change (1996), *Climate Change 1995, The Science of Climate Change*, Cambridge.

Kotzloff, Keith (1998), 'Electricity Sector Reform in Developing Countries: Implications for Renewable Energy', www.repp.org.

Kusch, Dietmar and Oliver Blank (1997), 'Die Situation der Energiewirtschaft', *Wirtschaftswelt China – Daten – Fakten – Analysen – Trends*, (5), 2–4.

Lile, Ronald, Mark Powell and Michael Toman, (1998), 'Implementing the Clean Development Mechanism: Lessons from U.S. Private-Sector Participation in Activities Implemented Jointly", Discussion Paper 99-08, November 1998, Washington, DC.

Meyer-Abich, Klaus Michael (1995), 'Neue Ziele – Neue Wege: Leitbild für den Aufbruch zu einer naturgemäße Wirtschaft und den Abschied vom Energiewachstum', in Enquete-Kommission 'Schutz der Erdathmosphäre' (ed.), *Mehr Zukunft für die Erde – Nachhaltige Energiepolitik für dauerhaften Klimaschutz*, Bundestagsdrucksache 12/8600, Bonn, pp.102-6.

Michaelowa, Axel (1997), *Internationale Kompensationsmöglichkeiten zur* CO_2 *– Reduktion*, Nomos, Baden-Baden.

Michaelowa, Axel and Holger Schmidt (1997), 'A dynamic crediting regime for Joint Implementation to foster innovation in the long term', *Mitigation and Adaptation Strategies for Global Change,* (2), 45–56.

Milborrow, David (1997), 'Wind is now cheaper than gas', in Rick Watson (ed.), *EWEC 97 – European Wind Energy Conference – Book of Abstracts*, Dublin, p.21.

Milborrow, David (1998), 'Renewables and the real world', *Windpower Monthly*, April, p.43.

Miljø og Energi Ministeriet (1996), Energy 21–The Danish Government's Action Plan for Energy 1996, Copenhagen.

Preuß, Olaf (1997), 'Chinas schmutziger Boom', *Greenpeace Magazin,* Heft 3/97, pp.16–23.

Solar Electric Light Company (SELCO) (1998), *Power for the 21st Century*, Chevy Chase.

Tatsis, Tass (1996), 'Key Considerations in Formulating Effective Market Mechanisms for the Commercialization of RETs', in Centro para a Conservação de Energia (ed.), *Market Incentives to Renewable Energies*, Seminar Proceedings, Lisbon, pp.27–42.

United Nations Development Programme (UNDP) (1997), *Energy after Rio – Prospects and Challenges*, New York.

8. Outlook on Climate and Development Policies

Axel Michaelowa and Michael Dutschke

HOW TO INDUCE DEVELOPING COUNTRIES TO ACCEPT FLEXIBLE INSTRUMENTS

Most of the developing world has been seeing flexible instruments as a pretext for industrialized nations to go on burning the earth's fossil fuel reserves while denying the developing world the benefits the industrialized countries have enjoyed until now. Others do not see any need at all for climate policy and think it diverts interest from development matters. The fact is that there are still doubts on the magnitude of the damage caused by greenhouse gases in the atmosphere, while the dangers of poverty, social inequity and pollution are proved daily.

However, there is a high degree of probability that today's developing countries will suffer most from climate change. The tropics are highly affected by disastrous climatic irregularities, like hurricanes, floods and droughts, that are likely to grow in number and magnitude.

Equity is an issue as well for the present and future distribution of emission rights. Development indicators being positively linked to GHG emissions, it is only natural that developing nations feel deprived of their right to development by the demand to limit their emissions. This is one reason why industrialized countries should go ahead and demonstrate that wealth can be achieved and maintained while decreasing GHG emissions (Engelmann, 1998). Only then will the catchphrase of 'technological leapfrogging', as illustrated in Chapter 7, become a credible option. Consumption patterns can only change globally, because the national elites of developing countries usually copy the industrialized countries' life style. In Kyoto emission rights were 'grandfathered' among the developed nations according to historical emissions and negotiation power. If in future commitment periods developing countries are expected to commit themselves to meaningful contributions, other distribution modes will be needed to determine each country's contribution to worldwide GHG emissions. Among these, there should certainly be a per capita ratio for emission rights. Or, to put it the other way around: just because a nation does not need its fair share of emission rights it should not be precluded from trading with these rights.

Any solution to the dilemma over emissions reduction and development has to address all the legitimate interests involved. In international climate negotiations, all the different actors have their hidden agenda as well. While, for instance, the power industry seizes the opportunity to promote its much-disputed nuclear plants, OPEC representatives try to obstruct the process in order to keep up their oil sales. Many representatives of the developing world see a chance for bargaining for new development aid. It must be reiterated that climate cooperation in itself has nothing to do with official development assistance (ODA). Theoretically, it should be an exchange on equal terms between unequal partners. This is why international conventions have to provide for a level playing field. This chapter recapitulates the major benefits induced for either side.

Thus, in order to make the flexible instruments acceptable to the developing countries which have rejected them so far, it is crucial that the industrialized countries take climate policy action at home, parallel to investing in JI, CDM and emission trading. They have an array of reduction measures at their disposal which are profitable for the national economy ('no-regret' measures) (see Figure 8.1) and which have not been implemented until now solely because of institutional impediments, lobbying activities and information deficits. These measures could not only result in a substantial emissions reduction, but also increase each nation's economic performance.

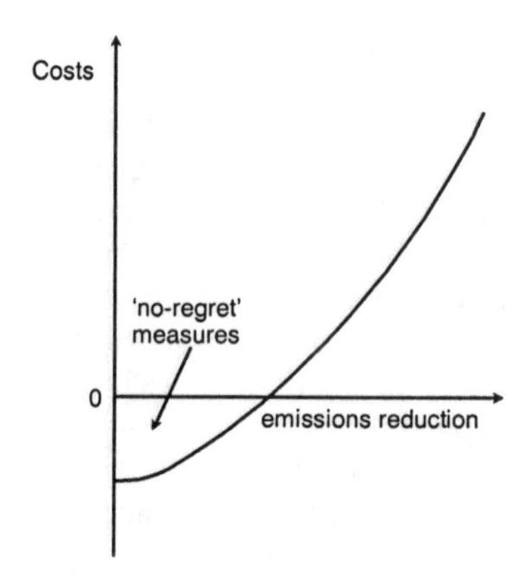

Figure 8.1 Economically profitable domestic emission reduction

Owing to political and economical factors, however, the implementation of these measures is extremely difficult, since it involves an intrusion in a web of protective measures which has evolved over a period of decades around politically sensitive branches of industry with powerful lobbies. Dismantling hard coal subsidies in Germany and other European countries, for example, would eliminate associated distortions and benefit the national economy to the tune of several billions.

If however, the industrialized countries are striving for a broad application of the flexible instruments they must go beyond 'no-regret' measures and, indeed, are virtually compelled to introduce tangible instruments such as a greenhouse gas tax with substantial rates of taxation. In such a context, the CDM would provide an opportunity to lead the developing countries, which even today cause more than half of the world's greenhouse gas emissions if land use change is included (Center for Clean Air Policy, 1998, p.2), on a path of climatically acceptable economic growth from the start (see Figure 8.2).[1] Many low-cost methods of reducing greenhouse gases cannot currently be used in the developing countries and the countries in transition because of financial bottlenecks (Trexler and Kosloff, 1993, p.3). The procurement of profitable technology often fails for lack of foreign exchange reserves. The financial mechanism of the Framework Convention on Climate Change do not provide sufficient capital. Therefore liquidity should be enhanced by offering incentives for private sector investment. Furthermore, successful CDM projects in the developing world can increase awareness of the climate issue and strengthen domestic efforts in these countries (Parikh, 1994, p.15). Developing countries can avoid costs for the future transition from an energy-intensive growth based on fossil fuels to a climatically acceptable economy, which sooner or later will be inevitable for them (Loske and Oberthür, 1994, p.5). At the same time, the flexible instruments will reinforce the massive structural change necessary in the industrialized countries by reducing transitional costs and allowing time for adjustment. Very high-cost measures can be postponed. In this way, structural change can be achieved smoothly and without severe adjustment crises.

1. In all of China, the fossil fuel power generation capacity is to be increased by 10GW p.a., hydroelectric power capacity by only 5GW. Coal's share of electricity generation is to held almost constant at 73 per cent (Li and Chen, 1994, p.720). CDM projects could lead to an increase in efficiency at the new power stations and push forward replacement of carbon-rich fuels. Coal-fired boiler efficiency could be increased by over 30 per cent nationwide by supporting such simple measures as the use of coal bricks instead of raw coal (Heidelberg Conference, 1994, pp.53f).

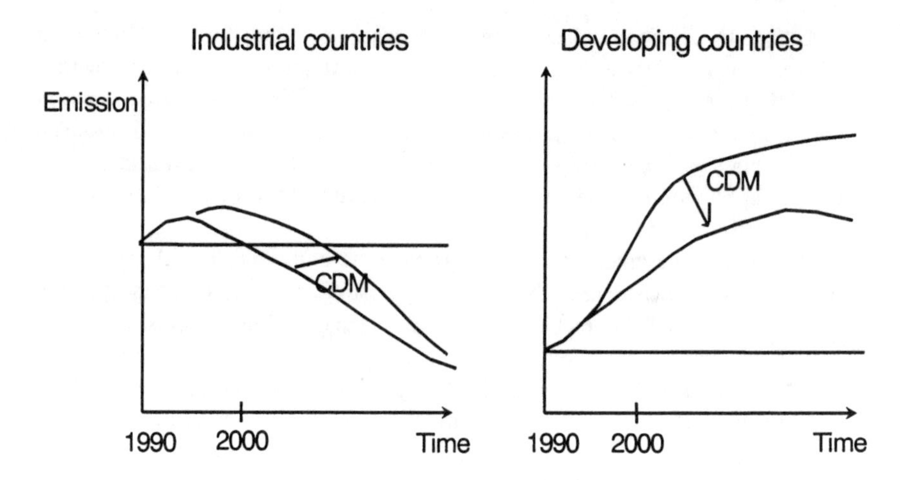

Figure 8.2 Cushioning structural change in the industrialized countries by increasing efficiency in the developing countries through the CDM

Emission reductions in the industrial countries can be spread over a longer time period while the emission efficiency of production growth in the developing countries is substantially improved. Given unchanged worldwide emission reduction costs, the CDM facilitates a higher global emissions reduction than is possible under climate policy which is restricted to the industrialized countries. If the political aim is to achieve the same emissions reduction with CDM as without it, then the costs of this reduction will be lower. The question of whether to aim for a higher level of emissions reduction at the same cost or the same level of emissions reduction at reduced cost requires a political solution.

SOLVING CRITICAL ISSUES IN THE DESIGN OF THE CDM

The CDM can become a vehicle to combine climate and development goals if negotiators never leave both objectives out of sight. A fair solution requires the determination of an institutional structure, baseline scenarios and how to share benefits resulting from climate cooperation.

Institutional Structure

The initiators of the CDM proposal clearly envisaged a multilateral fund. Given a multilateral solution, credits would accrue to the CDM, which distributes them to investors according to their share. The multilateral approach spreads project risks among all the investors, thus giving even conservative investors and investors with little capital a chance to participate. To raise funds for adaptation measures and administrative expenses, several possibilities exist:

- a part of credits created would accrue to the CDM and be sold on the market,
- a fixed percentage of the investors' payments could be deducted,
- the CDM could set a fixed price for credits and cover adaptation and administration costs out of the difference between project costs and the fixed price.

From a transparency point of view, the first solution would be preferable.

Besides operating as a fund, the CDM could also work as an international clearing house, operating in the same way as a broker or as pure project exchange. A CDM clearing house would accept and evaluate project proposals and invite possible investors to bid for projects (Hanisch, 1991; Mintzer, 1994, p.46). Credits accrue to the bidder who is succesful. Bidders would have to provide proof of insurance. Successful bidders would have to pay a charge for administration and adaptation purposes.

The leanest option for the CDM would be a project exchange where any interested party could gather quick, extensive information on all the climate projects currently available as well as on corresponding financial opportunities for funding the projects. The projects are all collected in an international database, access to which, via the Internet, is free of charge (see Mintzer, 1994, p.46, who gave this model the nice name of 'Hackers' Delight'). A fee is paid by the participants for successful matching to cover costs and raise adaptation project funds.

In any institutional structure, the CDM could provide a central or standardized insurance against the financial risk of failed JI projects. It should nevertheless differentiate its premiums according to the kind of project.[2] This insurance could be financed by retaining part of the credits and selling them on the market. Despite higher administration costs, a central insurance system can be more efficient for the individual contract partners than decentralized insurance. By spreading the insurance risk across all

2. This seems important, in order to prevent externalizing the high risks of forestry projects, for example, to other projects which are more expensive but carry lower risks.

climate cooperation projects and by standardizing procedural analysis, cost reductions can be achieved which will probably lead to lower premiums than could be offered by an individual project insurance. On the other hand, lack of competition could result in inefficiency and pure economic profits for the monopolistic central insurer.

In our view, it would be preferable to allow all types of institutional structures to compete on the basis of minimum rules set by the executive board. As the pure information exchange will not be palatable to the sceptics from the developing world and NGOs, only the first two models are feasible. From an economic point of view, it would be preferable to use them simultaneously as each has advantages for certain constituencies. Small investors will prefer the fund as they are not able to invest in a whole project. Moreover, their risk is lowered through the portfolio effect. Big investors will prefer to invest in whole projects as they can have synergy with other interests such as market development or technology transfer.

Determination of the Baseline

From a systematic point of view, the CDM is a loophole because it allows the industrial countries to inflate their cumulative targets. It is imperative to avoid this and ensure real and measurable emission reductions. The amount of emission reduction, obviously, depends on the emissions that would have occurred without the project: the 'baseline' of the project (Pearce, 1995, p.27). Obviously, cheating will be widespread if baseline determination is not subject to clear and strict rules. The longer the project's duration, the higher the uncertainties about its baseline. If the CDM host countries had quantitative targets themselves, there would be no need for baselines. Allowing the sale of credits on the international market for emission permits requires firm baselines and puts strict requirements on certifiers.

Often there is confusion about the differences between the definitions of 'additionality' and 'baseline'. We define the baseline as the overarching concept. The determination of whether a project is additional or not comes from calculating the difference between the verified emission of the project and the baseline emission. If the latter is higher, the project is 'additional'.

A major obstacle to defining a baseline is that the emission levels have to be forecast for the entire lifetime of the related project. In the case of carbon sequestration projects, the lifetime can be up to a century. Forecasting emission levels for such a long period amounts to guesswork. But, even for short-term projects (lasting five to ten years), it seems impossible to calculate an accurate baseline. The difficulty of business cycle forecasting is well known. Structural shocks can wreak havoc with a forecast: take, for example, an Eastern Europe development forecast in 1988 or an East Asian growth forecast of 1996. The question is, on the other hand, whether requirements for the CDM baseline have to be higher than for the Kyoto targets. These targets, fixed in December 1997 for emissions in the years 2008–12, are themselves based on a high degree of uncertainty.

The problems in establishing country-related baselines have already been felt in the business-as-usual projections in the national reports under the UN Framework Convention. Country baselines are necessary for determining reduction targets within the framework of international negotiations. Substantial evidence can be found that countries tend to overstate business-as-usual emissions, which can be used to negotiate from a position that offers a high reduction from the spurious baseline (Jochem *et al.,* 1994). If realistic baselines cannot be established, not only the CDM, but any other form of controlled greenhouse gas reduction policy, becomes impossible. Thus Heller (1998, p.12) argues for baselines that prevent the prolonging of inefficient economic structures. Cheating by individual project participants would become difficult if country-related baselines were used; but in that case, the host country government could try to set the parameters in a way that amounted to political distortion – that is, cheating on a macro scale.

Deliberate overstating of emission reduction would become rather difficult if a single standardized methodology for designing forecasting models and collecting data was required, to be drawn up by the Subsidiary Body on Scientific and Technological Advice (SBSTA) of the Convention (for a first preliminary collection of guidelines, see UNFCCC, 1997, pp.2–7). Baseline development should be subject to the review of independent certifiers: any creation of credits would take place only after examination of the baseline. It is also important to guarantee transparency and NGO participation in the process of setting baselines (Chomitz, 1998, pp.53ff). The SBSTA states that 'all AIJ require project-specific baselines. The methodologies used in calculating the baseline scenario may be sector-specific, technology-specific or country-specific' (UNFCCC, 1997, p.2).

Even if there is no deliberate attempt to cheat, different types of uncertainties exist concerning the baseline:

- political uncertainties, for example whether subsidies will be phased out;
- economic uncertainties, for example whether good policies put the country on a higher growth path or an external economic shock occurs;
- technological uncertainties, for example which technologies might have been chosen without the project;
- cost uncertainties, for example whether the project is a 'no-regrets' project.

One has to take into account that these uncertainties occur on different levels of aggregation: while the first two are primarily relevant on a country level, the third relates to the sectoral level and the fourth to the project level. But the first two can have effects on the project level too, depending on the degree of capacity utilization. It is impossible to develop baselines with no uncertainty whatsoever. However, we can evaluate them on the basis of whether they have a higher inherent uncertainty than a domestic climate policy.

Treatment of 'no-regrets' projects

The economic additionality of a project – determining whether it has positive costs compared with a commercially attractive alternative – is the most difficult issue in the context of baseline determination and has led to a heated debate (see Baumert, 1999; Rolfe, 1998 for an overview). Formerly, the whole baseline debate started from the definition of additionality (Carter, 1997).

Additionality can be seen on two levels: the country and the company level. Because of indirect effects, they will differ. A project that is clearly additional for a company may not be additional for a country as a whole. Under fossil fuel subsidies, for example, a wind power plant might be clearly additional because of higher costs compared with the subsidized fossil fuel. If the subsidy was phased out, it could become non-additional. Thus non-additionality at a country level will enhance the supply of projects that are additional at a company level.

Moreover, despite the theoretical profitability of many options, they face regulatory and legal obstacles, or are encumbered by a lack of information or skilled personnel, or by organizational rigidities. It is often reported that managers do not invest in promoting energy efficiency, even if its internal rate of return is much higher than the prevailing market interest rate. The main reasons are probably short planning periods, requirements for a minimal rate of return much higher than market interest and lack of capital. Another factor is the investor's planning security as far as political and fiscal conditions are concerned. It is not surprising that private households have even higher thresholds for internal rates of return. In particular, this applies to countries in transition and developing countries. Often an investor cannot appropriate a gain as it is an externality accruing to others. Therefore it seems that 'no-regrets' opportunities at the company level are quite scarce, whereas on a country scale there are many. In a similar vein, Heller (1998, pp.11ff) contends that transaction costs often inhibit 'no-regrets' projects.

Additionality at the company level could in theory be measured according to a number of criteria (see Figure 8.3). They assume that the discount rate and the degree of risk are known, which allows a calculation of risk-neutral costs.[3] (For a good discussion of the effect of different discount rates see Varming *et al.*, 1998).

1. Accept all projects that reduce or sequester emissions (as argued for by most of the business community and succinctly stated by Rentz, 1998).
2. Prove that the project removes barriers. A list of 'accepted' barriers could be defined (International Energy Agency, 1997).
3. Prove that the internal rate of return (IRR) of the project is lower than that of a commercial alternative.
4. Prove positive 'incremental' cost of the emission reduction-related part of the project similarly to procedures used by the Global Environment Facility. The loosening of these procedures in 1998 shows that they have been extremely difficult to apply.
5. Prove positive costs of the full project (for example by investment modelling) (Bedi, 1994).

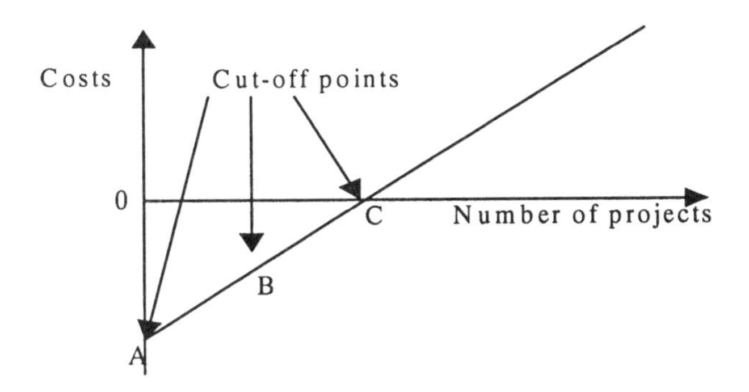

Notes:

A: All projects that reduce greenhouse gas emissions compared with status quo.

B: All projects above a negative cost threshold to account for non-monetary barriers and positive alternative rates of return.

C: Only positive cost projects are accepted

Figure 8.3 Determination of the economic additionality of a project

If cost data do not exist (for example, for confidentiality reasons) the following approaches might be helpful:

6. Use of a behavioural model (Chomitz, 1998).
7. Definition of project categories that are 'a priori' additional (Luhmann *et al.*, 1997).

---

3. Obviously, this is difficult to achieve as discount rates will be different from country to country and perceptions of risk are highly subjective.

Determining additionality at a company level may be impossible owing to the possibility of overstating cost figures (Torvanger *et al.*, 1994, p.21). A narrow additionality definition might lead to the choice of marginal technologies that are not always appropriate and depend on having an undistorted market, which often is not the case in host countries. Experience from AIJ shows that autonomous technology shifts depend on hard-to-observe parameters. On the other hand, country-level additionality might be easier to assess. Such an assessment would also have the advantage that there are no perverse incentives to prolong inefficient policies. The best approach would be to phase in strong country-related additionality rules over a certain period of time, such as five years, to allow countries to change policies. In that period, all projects that prove a greenhouse gas reduction from status quo for a period of five years should be accepted to account for higher transaction cost and barriers in the set-up phase of the CDM. That would kick-start the CDM and reward those businesses that have already started plans for less emission-intensive projects. Heller (1998, pp.15ff) even goes further, arguing that that a qualitative screening for transaction costs and barriers should be a sufficient criterion for key projects that prevent lock-in of energy-intensive infrastructures on a substantial scale to qualify for CDM.

Already many emission reduction projects are profitable to undertake – either for a company or for a country as a whole. The profitability for the latter includes externalities, such as the reduction of other pollutants. Now the question arises as to whether these so-called micro- or macroeconomic 'no-regrets' projects – projects that would seem to be profitable on their own – should be included in the baseline. So far, the question of 'no-regrets' has led to heated debates in the economics community. Some say that there can be no 'no-regrets' projects because such opportunities would have already been exploited (Sutherland, 1996). Others estimate that 10–30 per cent of today's emissions could be reduced by means of 'no-regrets' projects (IPCC, 1996).

Until now, the 'no-regrets' issue has been typically referred to in very general terms in the debate on the CDM and AIJ. While some authors say that all 'no-regrets' projects have to be included in the baseline and therefore excluded under the CDM (Bedi, 1994), others would accept all of them (for example, Rentz, 1998). Chomitz (1998, pp.3ff) presents a macroeconomic analysis of their inclusion. In this analysis, overstated baselines result in increased emissions, reduce the gains from CDM and divert rents away from projects with positive costs.

If microeconomic 'no-regrets' projects were excluded from the CDM, an investor would have an incentive to raise costs artificially to demonstrate that his project has positive net costs (Torvanger *et al.,* 1994, p.21). Therefore even that distinction cannot be applied. The International Energy Agency

(1997) tries to define barriers to project implementation. It suggests that a project be accepted under the CDM if it promises to overcome technological, financial or institutional barriers. This is the best proposal to evaluate the 'no-regrets' issue, but even here one can define barriers quite arbitrarily. Chomitz (1998) tries to quantify the barriers through behavioural and financial modelling. He states, however, that the parameters from which to create these models are known only to the project sponsor and can easily be manipulated. Thus he argues for default specification of these parameters on a country-by-country basis.

Constant versus revised baselines

The problems created by fixing baselines for projects with long lifetimes could be alleviated through revising baseline calculations regularly to take policy and macroeconomic changes into account (Andrasko *et al.,* 1996). This would result in increased uncertainty for the investors, because the credited emission reduction would depend on the adjustments of the baseline (UNFCCC, 1997, p.3). However, baseline regulations should limit, but not abolish, the entrepreneurial risks (and chances!) the investor takes. Emission reduction benefits should be useful for overcoming initial investment barriers, but they must not serve as a subsidy for an otherwise unprofitable project over its entire lifetime. Revising baselines could even fulfil the developing countries' demand for the transfer of state-of-the-art technology, by discouraging 'dumping' of outdated technology on the host country. If, for instance, a CDM project consists of the replacement of a coal-fired power station by an efficient renewable plant, the difference will not only be higher, but also longer lasting, than just retrofitting the old plant.

A possible compromise would be regular updating of the baseline – once every five years, for instance – or the option of a fixed, but more conservatively calculated, baseline (Hagler-Bailly, 1998, pp.1–9). The maximum lifetime of a baseline could be set at a decade. While annually updated baselines could be rewarded by certifying 100 per cent of the emission reductions, this portion could decline for projects with longer updating periods. In the second year of use of a baseline, for instance, 95 per cent of the emissions reductions would still be recognized as certifiable. Following this example, in year 10, only 55 per cent of the emission reductions calculated against the old baseline would be certified. Starting with year 11, a new baseline would have to be established for the project.

Countrywide baselines

An ideal country baseline would be an overall emission cap, comparable to the industrial country targets. Intra-country leakage would be zero and the 'no-regrets' issue sidelined. Nevertheless this is not politically feasible, because host countries are wary of committing themselves to anything like a cap. It still seems very appealing to calculate a baseline for a whole country and then aggregate the effects of the different CDM projects (Rentz *et al.*, 1998). For CDM host countries, we would expect a growth baseline. Reliable, quantified measurements of *actual* emissions are an important prerequisite for establishing such a baseline. A number of very different approaches are currently used to this end, producing highly divergent results. Hamwey (1997) simply averages historical emission factors. The study by Rentz *et al.* (1998) used energy systems models to derive baselines for Russia and Indonesia. However, because of the lack of data and data reliability problems, the Russian model does not consider industry and the household sector (ibid., p.160): it covers only the energy supply and forestry sector. Obviously, such a baseline will be misleading. The Indonesian model at least covers the electricity demand side, but not transport fuel or heat demand (ibid., pp.184ff). Countrywide baselines could be necessary if macroeconomic reforms such as subsidy phase-out were allowed to count as 'projects' under the CDM (Center for Clean Air Policy, 1998).

Sectoral baselines

A growing strand of literature (Carter, 1997; IEA, 1997) proposes sector-specific baselines. This conceptually simple, politically difficult solution would establish sectoral or national caps, and measure offsets against these (Carter, 1997). This is particularly appealing for large-scale projects with significant sectoral effects. For instance, a decision to build a generating plant can affect gridwide expansion and generation plans. Similarly, project-based efforts to protect particular forest plots from subsistence-oriented conversion may merely divert the farmers to another location. For large-scale energy and forestry projects, it could be desirable to compute sectoral level baselines and look at sectoral level effects. Sectoral baselines would have to be developed beforehand by the host country's institutions. This activity could be financed by the Global Environment Facility (GEF), thereby lowering transaction costs for potential investors.

Among the several severe difficulties of pursuing this approach is establishing the overall cap. This could be accomplished through the use of a complex model of the energy sector or of land use, based on prior emissions levels, adjusted for population or economic growth. In general, agreement on such a cap might be very difficult. A second difficulty arises in allocating the rights to create offsets against this cap. Moreover, the informational requirements can be very high, especially in a developing country context. Finally, it is unclear where sector boundaries should be set. Nevertheless, the sectoral approach should be chosen in the context of sink projects where leakage through simple relocation of forest destruction or degradation is very likely.

Benchmarks

Benchmarks (Hagler-Bailly, 1998) are quantitative emission factors per unit of output such as 'CO_2 emission per kWh of electric energy'. This would not limit the absolute rise of emissions and would therefore address developing countries' fears of being hampered in their economic growth. The advantage of benchmarks is highlighted by the extreme case of a power plant that is built but never operated: it would not yield any credits. Using benchmarks, increased sector production would lead to higher credits for the investing country only if the plant increased the overall efficiency. As high energy efficiency is a precondition for limiting CO_2 emissions, benchmarks could be used as common (and perhaps voluntary) targets for all parties to the Convention. However, this condition is necessary but not sufficient. The whole industrialization process has been marked by an increase in resource efficiency, which led to higher profits instead of diminished pressure on the resources. It would be counterproductive if industrialized countries were allowed higher absolute emissions while non-Annex countries increased their absolute emissions as well. Apart from damaging the atmosphere, this would erode the value of credits. Therefore mixing qualitative and quantitative targets in the long term does not lead to the desired results, but can be an intermediate strategy to integrate developing countries in the climate policy regime.

Nevertheless, benchmarks would motivate host countries to press for higher efficiency standards when accepting a CDM project. With the absolute emission reduction credits going to the investing country, the host country would profit by attaining its qualitative goal.

Project-related baselines

Taking into account the uncertainties of country-related baselines, *project-specific* baselines have been proposed as an alternative (Michaelowa, 1995, pp.65ff). The calculation of the baseline has to account for likely changes in relevant laws and regulations, the overall trend for efficiency improvements, and changes of other basic variables, such as development of markets for products of the project. It is possible to define either a 'median' baseline or a set of baselines with different assumptions weighted according to their probability (Andrasko *et al.,* 1996). For example, if a power station project does not replace existing plant but creates additional capacity, the baseline depends on the fuel that could have been used in an alternative solution. The alternative to a hydroelectric power station can be a coal-fired power station, for instance, burning either hard coal or lignite and producing very different emissions. For practical reasons, the host country's average fuel mix should be chosen for calculating the baseline in such cases (Michaelowa, 1995, p.65). The choice of such benchmarks is discussed by Hagler-Bailly (1998, pp.3–11ff), who discusses four ways of defining benchmarks: historical, forward-looking, and small and large samples.

The problem of defining an alternative project does not arise if, for example, an existing plant is to be replaced. In that case, the question of the remaining lifetime of the replaced plant has to be answered. Chomitz (1998, pp.6ff) lists a number of parameters that influence this value and finds them hard to observe, and subject to misrepresentation, strategic manipulation and autonomous change. Before the quantitative impact of a sequestration project can be estimated, relevant sources and sinks of greenhouse gases must be identified. Moreover, a quantification of past emissions is necessary. Demand-side management (DSM) projects pose special challenges, as they rely particularly on behavioural parameters. Nevertheless, the experience with US DSM has led to valuable progress in determination of actual energy savings that can be transferred to baseline determination (ibid., pp.14ff, 39ff).

To correct the estimates for 'free-riders' (those who would have installed the subsidized measures anyway) evaluators often use survey instruments. Remarkably, a significant proportion of the respondents acknowledge that they would have adopted the measures without any incentives. Another approach uses control groups. For instance, if high-efficiency light bulbs were subsidized through a CDM project in one city but not in an otherwise completely comparable control city, monitoring the latter would provide baseline information about the spontaneous rate of adoption of the bulbs in the absence of incentives. However, valid control groups are difficult to find because valid statistical comparisons require a large sample size for modest changes in emissions to be detected. Moreover, in many cases, the project facility may be unusual, and it may be difficult to find a large enough or similar enough control group to permit these comparisons. The control group has to be ineligible for, and likely to be unaffected by, the project.

Leakage

Project-specific baselines do not take into account indirect effects that can arise, for instance, when the project no longer produces but buys goods whose production caused greenhouse gas emissions. Emissions can also be influenced by price effects. For example, when carbon-rich fuels are replaced by low-carbon fuels, the price of the former tends to fall while the price of the latter will tend to increase. This price effect, in turn, tends to stimulate greater use of carbon-rich fuels and to lead to an increase in emissions. Demand-side energy savings would also cause energy prices to fall.

Another indirect effect would be the alleviation of energy supply shortages in host countries (Heister and Stähler, 1994). This argument is static, however. If one assumes rising incomes in these countries, these shortages would be alleviated in any case without emission reduction policies. It is possible, though, that industrial countries could try to push strongly for the extension of electricity supply in developing countries to enhance their own export markets for power supply technology. In this case, even the supply of efficient, state-of-the-art technology would lead to additional emissions compared to a business-as-usual path. Nevertheless, the emissions from additional electricity use would certainly be at least partly offset by reduction of emissions from unsustainable fuel-wood collection.

Therefore, an indirect effect of CDM projects might be to raise emissions in the short term but to lower them in the long term. It is likely that the latter effect would be greater. However, these indirect effects can only partially cancel out the emission reduction achieved by a CDM project. The effects described above arise in any sort of climate protection projects and not just in the case of CDM. Moreover, improved access to modern technology through the CDM can contribute to emission reductions. The same applies if products of the project sequester greenhouse gases and replace energy-intensive goods. It is impossible to specify whether indirect effects will lead to more or less emission reduction than the project-specific baseline scenario suggests. Thus, in the case of undistorted markets, there is no systematic trend for project-specific baselines to show excessive emission reductions. This is not taken into account by SBSTA (UNFCCC, 1997, p.2), which states that 'system boundaries for AIJ projects should be appropriate to the scale and complexity of the activity, so as to incorporate consideration of possible leakage'.

Nevertheless, leakage could be taken into account by deducting a certain percentage of leakage by default from the certified emissions reduction, depending on the type of project. The penalty such an approach exacts from low-leakage projects could be reduced by allowing project participants to get up to the full amount of certified emission reduction units if they can prove that leakage is lower than the default specification.

Besides indirect effects, a problem with project-related baselines arises if the host country distorts fuel and electricity markets by granting production or consumption subsidies. A project-related baseline cannot take into account changes in these subsidies that would alter a country-related baseline. As tight public budgets and liberalization of energy markets lead to subsidy cuts, project-related baselines would show greater emission reduction because of the higher incentive to save energy when energy prices rise. Thus, after phase-out of subsidies, we would forecast a lower countrywide baseline. A solution to this problem could be to prescribe a combination of a countrywide baseline with project-specific ones, which would allow for adjustment of the latter if the subsidies were phased out. This combination should be used only in cases of high subsidies or market distortion. It should be taken into account, however, that such a solution would provide a disincentive to phase out subsidies, because the amount of credits would be positively linked to the amount of subsidies.

Project lifetime

SBSTA proposed that projects with equity financing should use engineering or operating lifetime of the project, whereas projects with debt financing should use amortization or depreciation lifetime of the project (UNFCCC, 1997, p.3). Choosing the operating lifetime could eventually lead to running an outdated plant only because it generates emission reduction units. The amortization lifetime is usually too short a period to make use of the equipment´s full benefits. We therefore propose the commercial lifetime of the project's hardware, an intermediate measure, in order to keep the investment certifiable as long as it remains commercially profitable.

Credit Sharing

CDM rules described above can be interpreted to allow the following creation, allocation and distribution of credits, assuming credits are tradable according to Article 3.12 (see Figure 8.4):

- allocating credits only after discounting (see below);
- allocating credits only after a share has been deducted and sold to finance adaptation projects and administrative expenses (Art. 12.8);
- allocated credits accrue to the investing country in full: if the interpretation of 'benefits' accruing to the host country (Art. 12.3a) means investment capital and project externalities only;
- allocated credits are shared between investing and host country: 'benefits' (Art. 12.3a) means a share of the credits;
- An interesting, but daunting, option would be to set the share of the investing country at zero if the host country finances the project on its own;
- allocated credits accrue to the investing country only until a quota is reached (Art. 12.3b).

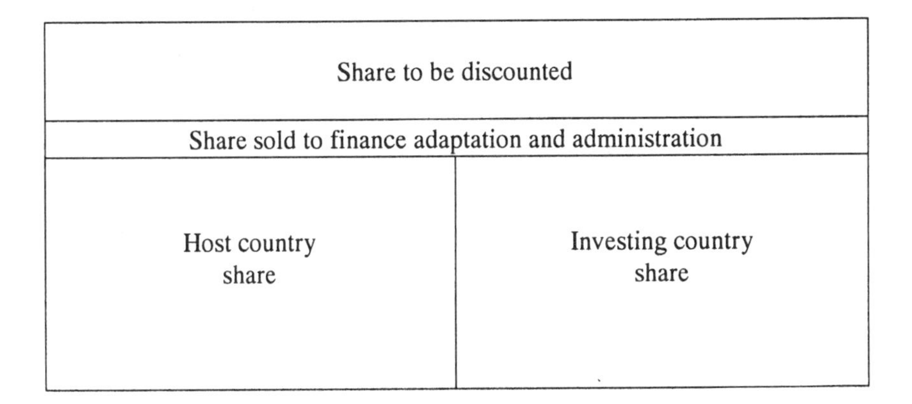

Figure 8.4 Possible credit allocation and distribution

CDM is often understood to involve the following type of transaction: a government or a private entity of a country with an emissions target finances a project in another, the host country. Credits from the emission reduction accrue only to the investor. The positive externalities from the project (see below) are deemed sufficient as incentive for the host country. If credits are fully tradable domestically and internationally, they should accrue to the entity investing, even if it is a private company. If there is no domestic trading system, the credits should accrue to the government which, in turn, should compensate the investor through emission tax reductions or reductions in regulatory requirements.

The host country will be interested in credits when one or more of the following applies: it is subject to an emissions target; it does not have an emissions target now but wants to bank credits for future commitments; credits can be traded on a market.

Allocating all credits to the host country would make no sense for a rational CDM investor. Of course, the host country could then finance the project on its own and sell credits earned. No rule of Article 12 would prevent this. Costa Rica has already pioneered this kind of trade by financing umbrella forestry and energy projects through a fuel tax and trying to sell certified tradable offsets (see Chapter 6).

Such a general participation of host countries in creating and trading credits would certainly lower the price of credits and ring alarm bells in many quarters, especially if credits could be traded from 2000 onwards. As host countries have no targets, they have an incentive to maximize credit sales. Here the baseline issue becomes crucial: the situation has to be avoided where there is a reward for developing countries if their policy promotes high emissions. This is due to a perverse effect of the additionality rule: emission reduction measures are cheapest where there is a lack of a national sustainability policy (Michaelowa and Dutschke, 1997, p.46). The CDM would have to be extremely cautious concerning baseline verification.

This problem could only be fully solved by setting an incentive for developing countries to adopt limitation targets and participate in JI and emissions trading under Articles 17 and 6. Such an incentive could be to prohibit the trading of host country credits now but to allow them to bank credits against future targets. One could also envisage a quota for credit trades for each country and banking for additional credits created.

The national emission targets should be derived from national baselines developed using common rules and procedures. Any improvement in environmental legislation will then be beneficial for future compliance. As in the case of the investing country, credits could either accrue to the entity involved in the project or the government. The former would be only relevant if credits could be traded freely. The decision on that issue could have important distributional consequences. To sum up: credit sharing leads to higher costs for the investors. Free negotiation of the credit sharing ratio will lead to competition between host countries.

CREATING SYNERGIES: SUCCESSFUL INTERTWINING OF CLIMATE AND DEVELOPMENT GOALS

Positive externalities are the only incentive for host countries to engage in CDM unless they receive a part of the credits and are able to sell them. The following externalities are relevant:

- formation of human capital,
- transfer of technology,
- capital transfer,
- foreign currency transfer,
- job creation,
- improvement of distribution,
- reduction of local pollutants,
- protection of biodiversity.

It is very difficult to quantify these externalities. Most of them are interlinked and operate on different time scales. Feedback effects depend on the local situation. While it is obvious that CDM projects will lead to capital and foreign currency transfer, the net effects on jobs are unclear. The transfer of modern technology could well lead to a loss of jobs, at least locally and in the short and medium term. Formation of human capital is a long-term effect and dependent on the social and political framework. Improvement of distribution also depends on the local political and social situation.

Tentative calculations (Ekins, 1996) show that the benefits of emission reduction through reduction of local pollutants, especially SO_2, are comparable to the value of carbon credits under a high carbon tax of US$20–200 per ton of carbon. Thus the value of externalities of carbon emission reduction would in fact be higher than the credit value from CDM projects accruing to the emitter under a moderate domestic climate policy regime. As the critical loads of local pollutants have not yet been reached in many developing countries, the benefit stemming from carbon emission reduction would be lower compared with industrialized countries. Nevertheless, it seems that reduction of local pollutants will be a relevant externality particularly for densely populated countries in transition and newly industrializing countries, for example in Asia. Projects which offer such benefits will be preferred. In fact, the first AIJ project in central Europe, the Decin project in the Czech Republic, was promoted by the local authorities for precisely that reason.

Biodiversity will only be protected if the social and political framework is conducive to forest protection and prevents relocation of damaging activities. Thus only countries with a strong administrative capacity are able to take advantage of biodiversity-related CDM. Costa Rica is an example of such a trend, as it focused especially on extension of national parks through AIJ funds. It is likely that the capital and technology transfer will be decisive for those host countries where official development aid is declining. Countries with high private capital flows will try to use CDM funds to maximize positive environmental and social externalities.

It is crucial to minimize negative externalities. The most critical negative externality of climate cooperation projects could be that they reduce incentives for innovation. For a detailed discussion of this aspect, see Michaelowa and Schmidt (1997). It would be advisable not to set quotas on the share of CDM credited towards industrial countries' targets as they give no dynamic incentive for innovation.

Other negative externalities could include displacement of people and loss of arable land in the case of large-scale hydro and afforestation projects. Many negative externalities are linked to poor management and an unstable political situation. It is probable that many projects will have a mixture of positive and negative externalities. The question of how to weight them will be crucial for the success of these projects.

Crediting and Externalities

From the preceding discussion it is obvious that there is no general rule for the sign and size of externalities. An exact quantification is impossible and the situation is different for each project. Because of high transaction costs, it is not advisable to calculate externalities for each project. Nevertheless, certain project types are more likely to entail positive externalities than others. Fossil power plants will create fewer jobs than demand-side management programmes. Renewable energy projects will mean zero emission of local pollutants, compared with fuel substitution projects that still lead to a – however reduced – emission. Forest protection projects and afforestation are unlikely to entail technology transfer and human capital formation. The former are likely to entail biodiversity protection, while the latter is not. Large-scale projects are more likely to disrupt local life and displace people than small-scale ones. The following general conclusions can be used to categorize projects and differentiate crediting.

1. Demand-side management and production of renewable energy can be credited fully.
2. Large-scale projects such as new fossil power plants, but also forest protection, are only credited partially. The latter are included in this category despite their high biodiversity externality because of uncertainties concerning relocation of deforestation. Alternatively, these risks could be covered by a compulsory insurance.
3. Afforestation should be credited at a low rate as it rarely entails technology transfer and leads to land use constraints. The risks of reversal have to be covered adequately.

Moreover, the host country can grant incentives on its own. For example, in the case of private enterprise sequestration projects, state grants should help make diversified reforestation projects more attractive. This sort of procedure is beneficial for the economy as a whole, since more diversified projects involve greater positive externalities. The preservation of threatened areas of primary forest can bring even greater positive externalities.

Crediting of externalities should be explicit if there are other incentives such as subsidies for biodiversity protection not payable to CDM projects. Even if externalities were not credited explicitly, an investor would profit from a project with many positive externalities because approval and execution of the project would be smoother than otherwise.

Concerning innovation, on the one hand there have to be incentives for induced innovation to achieve long-term efficiency gains. On the other hand, short-term efficiency gains through CDM have to be allowed. A ‘strategic’ climate policy could entail a sliding reduction of exploitable short-term efficiency gains while raising an emission tax in the long run. Figure 8.5 shows a sliding reduction of crediting of CDM. In the same period, either domestic carbon taxes are raised with a steadily rising tax rate in the industrialized countries or a system of tradable permits with a steadily sinking supply is introduced.

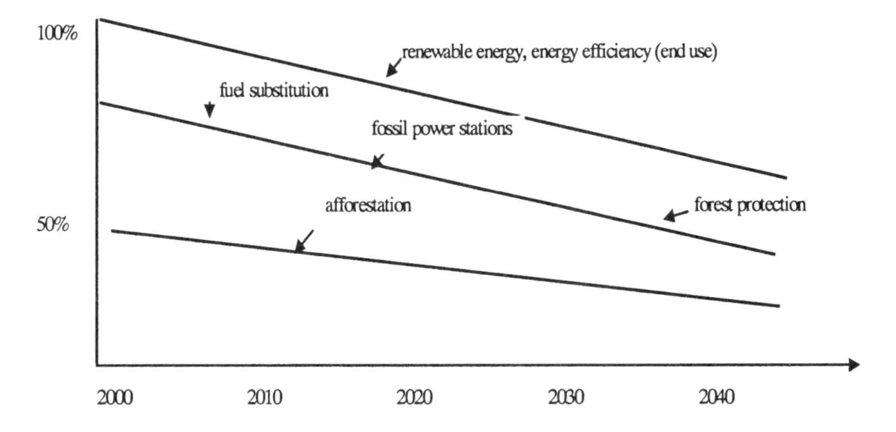

Figure 8.5 Decreasing credit ratios

This policy entails the following advantages:

- Investors receive long-term planning data.
- Investors can get full crediting for CDM in the beginning which allows them to invest into long-term emission reduction strategies. However, crediting is linked to the extent of technology transfer:
- The incentive to reduce domestic emission grows steadily as crediting falls, while the emission tax rises.

Towards Global Emission Trading on an Equitable Basis

In the long run, a global system of emission trading will be the most efficient way to reduce greenhouse gas emissions. Developing countries will only accept such a system if it takes into account equity. In general, an equitable climate policy should mean that each inhabitant of the world has equal rights to the atmosphere. In economic terms, that means that everybody would be entitled to the same quota. The target could also take into account historical emissions which have not yet decayed and therefore stress the long-term nature of the problem (Agarwal and Narain, 1991). To divide annual emissions into the number of people, calculating an average and thus allocating emission permits is straightforward. A global reduction target would then mean a proportional devaluation of every permit. As population changes are not tracked every year, reallocation should occur only every five years. There should be a mechanism to prevent inflation of population numbers such as an independent evaluating team. If climate policy is also used for population policy purposes, the population could be fixed at a date where the climate problem became politically relevant (for example, 1992; see Bohm, 1995, p.26). One could also choose the population before the onset of the demographic transition, but fixing the date by consensus will be difficult.

To redress historical inequities in emissions, cumulated emissions have to be taken into account. There could be different formulae concerning the cumulated emissions. A strict one would divide the cumulated emissions (minus natural decay) into the number of people living or having lived in the period. Then one could calculate a global average of cumulative emissions per capita and calculate emission debtors and creditors. Accordingly, the allocation of emission permits would be changed.

From a responsibility point of view, one could limit the calculation of past emissions to the year where the connection between greenhouse gas emissions and climate change became obvious beyond purely scientific circles. A suitable date would be the First World Climate Conference in 1979. For a calculation of different allocation modes, see Box 8.1.

Box 8.1 Equity-based allocation and reduction necessity per group of countries

The allowable global emission[4] of 6000Gt CO_2 must be distributed over time. The higher the peak emissions, the less natural decay can reduce the concentration and the higher the risk that a critical threshold is passed. The WBGU emission path (WBGU 1996) would mean that global emissions would be reduced from the current value of 22 300Mt by 1 per cent p.a., that is initially 223Mt. Compared to the business-as-usual rise of 1.7 per cent p.a., the necessary reduction will be a further 380Mt. It is assumed that the ocean sinks are treated as a global resource that does not belong to a particular country. How can the emissions be allocated equitably?

4. Gt: billion metric tons, Mt: million metric tons. Including other greenhouse gases would shift the balance in favour of the industrial countries as their emissions are more broadly distributed. As the database for emissions of those gases is still incomplete and unreliable, a calculation of the mitigation commitments is impossible. The principles explained above could be extended to these gases as soon as the data become more reliable.

1. Per capita allocation of future emissions, population 'frozen' at 1990 level

If the current emission is allocated equitably, each person has the right to emit 4.2t p.a. All industrial countries and NICs would have to reduce their emissions by a factor of two to five to reach that value. Most developing countries with the exception of some OPEC countries, do not reach the threshold. To reduce incentives for further population growth, the allocation should be fixed at 1990 population figures. To reach the threshold, industrial countries can at first use the CDM. Taking into account a 'no-regret' potential of 10–20 per cent in the industrial world, the OECD could reduce per capita emissions at home from currently 11.2t to 9.5t with economic gains. Then, annually about 5000Mt would have to be reduced either via CDM or at home. As the emissions of developing countries and countries in transition which currently are 10 000Mt, could not realistically be reduced by 50 per cent, bilateral contracts on the temporary lease of unused emission permits of developing countries could be concluded. At a later stage these emission permits could be made tradable and industrial countries could buy surplus permits from developing countries.

With an annual reduction of 1 per cent, global emissions would be below 21 500Mt in 2000, below 20 000Mt in 2007 and below 17 500Mt in 2020. By then, world population will have reached 8.1 billion and the real per capita allocation accordingly will have almost halved to 2.2t p.a., while, taking into account that the population shares are fixed at the 1990 levels, the 1990 weighted value would be 3.3t. Many developing countries will have crossed that threshold and the permit surplus will be limited to a few countries. In the long run, decay of CO_2 could lower the restrictions and allow a residual emission that has been quantified by WBGU at around 15Gt. This value would be reached by 2035.

2. Per capita allocation including accumulated emissions since 1980 (weighted with 1990 population)

Since 1980, global warming has been discussed widely, so every country has to bear responsibility for its emissions over this period. The OECD countries emitted more than 100Gt in the 1980s, accumulating a 'debt' of more than 100t per capita. A similar relation applies for the countries in transition. The corresponding value for the rest of the world is only a fourth and the global average about 42t per capita. It would be appropriate for countries which have emitted less than the global average to receive an additional allocation which is 'financed' by an additional reduction commitment of those countries which have overused the resource. The compensation should be extended over a period of several decades to prevent financial strain. If the OECD countries had to mitigate their additional share of 56Gt over a period of 40 years, their additional annual commitment would be 1.4Gt, or about 13 per cent of current OECD emissions. This would not be achievable without huge economic costs. Therefore a period of a century would be appropriate, that is, 0.6Gt per year (6 per cent of current emissions).

3. Per capita allocation including accumulated emissions since 1950 (weighted with 1990 population)

If the 'polluter pays principle' was accepted, emitters would have to be hold liable for all greenhouse gas emissions that are still in the atmosphere. Accordingly, emissions would have to be discounted by the natural decay factor and then cumulated. As the database for country-specific emissions extends only to 1950, the predating emissions cannot be assigned to each country. Furthermore, there are difficulties concerning the allocation of emissions of colonies: should they be allocated to the colonial power or the territory? A pragmatic approach would therefore entail a calculation of the cumulative emissions. The calculation below does not entail a decay factor and therefore is biased against those countries which had high emissions in the earlier part of the period. A territorial approach is chosen. If the global total of 557 050Gt is allocated to the population living in 1990, a global average of 105t per capita can be deduced. The OECD average would be 320t while the average of the rest of the world would be 55t. Allocating the debt according to (2) the additional commitments of the OECD would be 202Gt. The compensation should be extended over more than a century to keep annual mitigation rates below 2Gt.

From the moral point of view, the developing countries should press for an immediate per capita allocation. Because of the political and economic power of the industrial countries, it is unlikely that these will accept the equity-based allocation in the near future, unless well organized NGO activities worldwide lead to a climate conducive to strong redistribution. In the case of industrial country resistance, the climate policy debate could be compared with the debate on the New Economic Order in the 1970s. That was a time when a subset of developing countries (OPEC) could put real pressure on the industrial countries. Still, the debate ended in a dead end. Another example of high demands by the developing countries and an ensuing stalemate is the negotiations on the Law of the Sea which took over a quarter of a century before the Convention on the Law of the Sea finally entered into force – much watered down. Moreover, if time is lost in that way, the industrial countries will surely point to the rapidly rising emissions from the developing countries, thereby reducing allocation of permits and transfers. Despite the dismal record of technology transfer in aid or environmental agreements so far, there is a good chance that self-interest of the industrial countries will lead to a better outcome.

If the right incentives are offered to CDM investors, money and technology will start to flow in on a large scale. Developing countries should therefore press for a system of tradable permits that could start with an allocation not much different from the existing one, but with a well-defined path (such as a sliding 20 per cent reduction compared to 1990 of the industrial countries' permits until 2010, 50 per cent reduction until 2030, and 70 per cent until 2050) towards the per capita allocation.

Therefore developing countries should adopt a step-by-step approach. The first step is CDM, as it gives the developing countries implicit entitlements which can be exchanged against capital and technology transfers. This step should begin in 2000, as originally envisaged under the Kyoto Protocol. As a second step, national permit systems could be introduced before the global system of tradable permits is introduced. Of course, under the CDM transparency is lower and investment depends on the incentives granted by the governments of industrial countries. The amount of transfer to the developing countries also depends on the negotiating skill of developing country actors. From the discussion above, it has become obvious that there are many obstacles to an efficient, first-best application of the CDM in an international as well as a national context. A CDM regime which is in the hands of interest groups could lead to a number of distortions.

1. The emitters in the industrial countries use the regime to purport that they are doing something about climate change and to prevent strong domestic instruments of climate policy.
2. Emitters in industrial and developing countries join forces to claim spurious reductions.
3. Politicians in industrial countries want to reduce development aid and see the CDM as a valuable substitute.
4. Emitters in industrial countries lobby for less costly instruments. Subsidies and voluntary commitments are therefore more likely to be the basis for CDM than the more efficient emission taxes or tradable emission permits.
5. Emitter interests still being better organized than NGO interests, the gains of CDM projects might accrue disproportionally to the former. They will use the CDM to lower costs instead of achieving higher emission reductions.
6. Large projects are preferred by the emitters in the investing countries and the politicians of the host countries. NGO and local interests in small-scale projects with high positive externalities will be defeated.

In order to combat these distortions, pressure groups must be prevented from appropriating the CDM for their special interests. As in all political fields, this can only be guaranteed through a system of checks and balances, which, however, takes away some of the efficiency properties. Two main principles can be found which imply some restrictions on the application of the CDM. Therefore, in comparison with the efficiency-oriented solution, one can speak of a second-best solution. It is characterized in the following way: (a) through *transparency*, the actions of participants can be subjected to close scrutiny, distortions are less likely and the regime will gain credibility; (b) political action must be *rule-oriented* and should not be discretionary; interest groups find it more difficult to change rules than to change single decisions. If the rules are strong, policy makers cannot intervene easily. For a second-best application of the CDM these mechanisms should be established at the national and international level to counterbalance lobbying activities.

As early as possible, a global system of permits should be instituted that will be the most efficient, transparent and equitable way to reach the long-term target of stabilizing greenhouse gas concentrations. The permit price will give the same incentive to reduce emissions all over the world. The permits would be traded by governments, companies and individuals and form a kind of money. They would be issued by a global clearing house that steered the quantity of permits according to the optimal global emission path. To prevent political deadlocks, the clearing house should be operated by a council of directors according special voting rules. For example, they could specify that a majority of both permit creditor and debtor countries is required for important decisions. In monetary terms, an independent central bank has proved to be successful in stabilizing the value of currencies. Accordingly, a climate central bank could help enforce quantitative goals by limiting the absolute number of emission rights (Dutschke *et al.*, 1998). As long as developing countries stay outside the system, the 'climate bank' could administrate the aggregate of emission rights, in order not to inflate the system. Each emission reduction unit created outside the system (that is, by the CDM or by sequestration measures within committed countries) would be allocated free to the investor. On the other hand, it would have to be withdrawn proportionally from the aggregate of the committed countries' budget.

However, the allocation rule should operate on a per capita basis as soon as possible, in order to include developing countries. Taking into account the political realities, a gradual change from grandfathering to a per capita rule seems most promising. In the long run, the accumulated emissions debt would also have to be 'repaid' (see Box 8.1). To prevent an 'emission permit debt' problem which could be created if corrupt governments raise short-term revenue by selling the whole futures stock of emission permits, futures trades by surplus countries could be limited to a period of, for example, five years. The role of CDM will then have finished, as there is no further need to organize emission reduction projects on a bilateral or multilateral basis: they will be automatically induced by the permit price. The price depends on the emission targets and the allocation. Via futures markets, national development strategies could be secured against variations in the permit price.

REFERENCES

Agarwal, Anil and Sunita Narain (1991), *Global Warming in an Unequal World*, New Delhi.

Akumu, Grace (1993), 'Joint Implementation: Mitigation strategy or hidden agenda?', *Eco*, (4), 3.

Andrasko, Kenneth, Lisa Carter and Wytze van der Gaast (1996), 'Technical Issues in JI Projects', background paper for the UNEP AIJ Conference, Washington.

Baumert, Kevin (1999), 'Understanding additionality', in José Goldemberg and Walter Reid (eds), *Promoting Development while Limiting Greenhouse Gas Emissions: Trends and Baselines*, New York, pp.135–43.

Bedi, Chiranjeev (1994), 'No regrets under Joint Implementation?', in Prodipto Ghosh and Jyotsna Puri, (eds), *Joint Implementation of Climate Change Commitments*, New Delhi, pp.103–7.

Bohm, Peter (1995), 'An analytical approach to evaluating the net costs of a global system of tradable carbon emissions entitlements', UNCTAD/GID/9, Geneva.

Carter, Lisa (1997), 'Modalities for the operationalization of additionality', in UNEP, BMU (eds), *Proceedings of the International AIJ Workshop*, Leipzig, pp.79–92.

Center for Clean Air Policy (1998), *Top-down Baselines to Simplify Setting of Project Emission Baselines for JI and the CDM*, Washington.

Chomitz, Kenneth (1998), 'Baselines for greenhouse gas reductions: problems, precedents, solutions', World Bank Carbon Offset Unit, Washington.

Dutschke, Michael, Axel Michaelowa and Marcus Stronzik (1998), 'Tightening the system: Central allocation of emission rights', HWWA Discussion Paper no. 70, Hamburg.

Ekins, Paul (1996), 'How large a carbon tax is justified by the secondary benefits of CO_2 abatement', *Resource and Energy Economics*, (18) 161–87.

Engelmann, Robert (1998), 'Population, consumption and equity', *Tiempo*, 30 December, 3-10.

Hagler-Bailly (ed.) (1998), *Evaluation of Using Benchmarks to Satisfy the Additionality Criterion for Joint Implementation Projects*, Boulder.

Hamwey, Robert (1997), 'A sustainable framework for Joint Implementation', IAE Working Paper 51, Geneva.

Hanisch, Ted (1991), 'Joint implementation of commitments to curb climate change', CICERO Policy Note 1991:2, Oslo.

Heidelberg Conference (1994), 'How to combat global warming at the local level', *Compendium of Abstracts*, Heidelberg.

Heister, Johannes and Frank Stähler (1994), 'Globale Umweltpolitik und Joint Implementation: Eine ökonomische Analyse für die Volksrepublik China', *Kieler Arbeitspapier* no.644, Kiel.

Heller, Thomas (1998), 'Additionality, transactional barriers and the political economy of climate change', *Fondazione Eni Enrico Mattei Nota di lavoro*, 10.98, Venice.

Intergovernmental Panel on Climate Change (1996), *Climate Change 1995. The Economic and Social Dimensions of Climate Change*, Cambridge.

International Energy Agency (1997), *Activities Implemented Jointly–Partnerships for Climate and Development*, Paris.

Jochem, Eberhard, Heinrich Herz and Wilhelm Mannsbart (1994), *Analyse und Diskussion der jüngsten Energiebedarfsprognosen für die großen Industrienationen im Hinblick auf die Vermeidung von Treibhausgasen*, Bonn.

Li, Naihu and Heng Chen (1994), 'Umweltschutz in der elektrischen Energieversorgung Chinas', *Energiewirtschaftliche Tagesfragen*, 44(11), 718–25.

Loske, Reinhard and Sebastian Oberthür (1994), 'Joint Implementation under the Climate Change Convention: Opportunities and Pitfalls', *International Environmental Affairs*, (1), 45–58.

Luhmann, Hans-Jochen, Hermann Ott, Christiane Beuermann, Manfred Fischedick, Peter Hennicke and Liesbeth Bakker (1997), 'Joint Implementation – Projektsimulation und Organisation: Operationalisierung eines neuen Instruments der internationalen Klimapolitik', Berlin.

Michaelowa, Axel (1995), 'Joint Implementation of greenhouse gas reductions under consideration of fiscal and regulatory incentives', HWWA Report no. 153, Hamburg.

Michaelowa, Axel and Michael Dutschke (1997), 'Joint Implementation as development policy – the case of Costa Rica', HWWA Discussion Paper no. 49, Hamburg.

Michaelowa, Axel and Holger Schmidt (1997), 'A dynamic crediting regime for Joint Implementation to foster innovation in the long term', *Mitigation and Adaptation Strategies for Global Change*, (2) 45–56.

Mintzer, Irving (1994), 'Institutional options and operational challenges in the management of a Joint Implementation regime', in Kilaparti Ramakrishna (ed.), *Criteria for Joint Implementation under the Framework Convention on Climate Change*, Woods Hole Research Center, pp.41–50.

Parikh, Jyoti (1994), 'Role of markets, governments and international bodies in Joint Implementation with the South', in Kilaparti Ramakrishna (ed.), *Criteria for Joint Implementation under the Framework Convention on Climate Change*, Woods Hole Research Center, pp.15–24.

Pearce, David (1995), 'Joint Implementation: A general overview', in Catrinus Jepma (ed.), *The feasibility of Joint Implementation*, Dordrecht, pp.15–31.

Rentz, Henning (1998), 'Joint Implementation and the question of additionality - a proposal for a pragmatic approach to identify possible Joint Implementation projects', *Energy Policy*, (4), 275–79

Rentz, Otto, Martin Wietschel, Armin Ardone, Wolf Fichtner and Matthias Göbelt (1998), *Zur Effizienz einer länderübergreifenden Zusammenarbeit bei der Klimavorsorge*, Karlsruhe.

Rolfe, Chris (1998), *Additionality: What is it? Does it matter?*, Vancouver.

Sutherland, Ronald (1996), 'The economics of energy conservation policy', *Energy Policy*, 24(4), 361–70.

Torvanger, Asbjorn, Jan Fuglestvedt, Catrine Hagem, Lasse Ringius, Rolf Selrod and Hans Aaheim (1994), 'Joint Implementation under the Climate Convention: Phases, options and incentives', CICERO Report 1994:6, Oslo.

Trexler, Mark and Laura Kosloff (1993), 'US should encourage foreign offsets to meet year 2000 emissions goal', *Global Environmental Change Report*, (15), Supplement, 2–3.

UN Framework Convention on Climate Change (1997), 'Activities Implemented Jointly under the pilot phase', FCCC/SBSTA/1997/INF.3, Bonn.

Varming, Sören, Peter Larsen and Brigitte Christens (1998), 'Possibilities of AIJ: two Polish cases', in Pierce Riemer, Andrea Smith and Kelly Thambimuthu (eds), *Greenhouse Gas Migration. Technologies for Activities Implemeted Jointly*, Amsterdam, pp.529–34.

Wissenschaftlicher Beirat der Bundesregierung Globale Umweltveränderungen– (WBGU) (1996), *Welt im Wandel – Wege zur Lösung globaler Umweltprobleme, Jahresgutachten 1995*, Heidelberg.

Appendix

JOINT IMPLEMENTATION IN THE FRAMEWORK CONVENTION ON CLIMATE CHANGE

Article 3: Principles

[...]

Art. 3.3. The Parties should take precautionary measures to anticipate, prevent or minimise the causes of climate change and mitigate its adverse effects. Where there are threats of serious or irreversible damage, lack of full scientific certainty should not be used as a reason for postponing such measures, taking into account that policies and measures to deal with climate change should be cost-effective so as to ensure global benefits at the lowest possible cost. To achieve this, such policies and measures should take into account different socio-economic contexts, be comprehensive, cover all relevant sources, sinks and reservoirs of greenhouse gases and adaptation, and comprise all economic sectors. Efforts to address climate change may be carried out co-operatively by interested Parties.

Article 4: Commitments

[...]

4.2. The developed country Parties and other Parties included in Annex I commit themselves specifically as provided for in the following:

(a) Each of these Parties shall adopt national policies and take corresponding measures on the mitigation of climate change, by limiting its anthropogenic emissions of greenhouse gases and protecting and enhancing its greenhouse gas sinks and reservoirs. These policies and measures will demonstrate that developed countries are taking the lead in modifying longer-term trends in anthropogenic emissions consistent with the objective of the Convention, recognising that the return by the end of the present decade to earlier levels of anthropogenic emissions of carbon dioxide and other greenhouse gases not controlled by the Montreal Protocol would contribute to such modification, and taking into account the differences in these Parties' starting points and approaches, economic structures and resource bases, the need to maintain strong and sustainable economic growth, available technologies and other individual circumstances, as well as the need for equitable and appropriate contributions by each of these Parties to the global effort regarding that objective. These Parties may implement such policies and measures jointly with other Parties and may assist other Parties in contributing to the achievement of the objective of the Convention and, in particular, that of this subparagraph;

(b) In order to promote progress to this end, each of these Parties shall communicate, within six months of the entry into force of the Convention for it and periodically thereafter, and in accordance with Article 12, detailed information on its policies and measures referred to in subparagraph (a) above, as well as on its resulting projected anthropogenic emissions by sources and removals by sinks of greenhouse gases not controlled by the Montreal Protocol for the period referred to in subparagraph (a), with the aim of returning individually or jointly to their 1990 levels these anthropogenic emissions of carbon dioxide and other greenhouse gases not controlled by the Montreal Protocol. This information will be reviewed by the Conference of the Parties, at its first session and periodically thereafter, in accordance with Article 7;

[...]

(d) The Conference of the Parties shall, at its first session, review the adequacy of subparagraphs (a) and (b) above. Such review shall be carried out in the light of the best available scientific information and assessment on climate change and its impacts, as well as relevant technical, social and economic information. Based on this review, the Conference of the Parties shall take appropriate action, which may include the adoption of amendments to the commitments in subparagraphs (a) and (b) above. The Conference of the Parties, at its first session, shall also take decisions regarding criteria for joint implementation as indicated in subparagraph (a) above. A second review of subparagraphs (a) and (b) shall take place not later than 31 December 1998, and thereafter at regular intervals determined by the Conference of the Parties, until the objective of the Convention is met.

Tentative Criteria for Joint Implementation by the Interim Secretariat of the Framework Convention (Prepared for the 9th Session of the INC)

1. Joint implementation refers only to joint action to implement policies and measures, and in no way modifies the commitments of each Party.
2. Joint implementation is distinct from the provision of assistance to other Parties.

3. Joint implementation is a voluntary activity under the responsibility of two or more Parties; such activity must be undertaken or accepted by the Governments concerned.
4. Joint implementation would be undertaken in conjunction with domestic action.
5. Joint implementation should be beneficial to all Parties involved, and be consistent with their national priorities for sustainable development.
6. Joint implementation activities should bring about real and measurable results, determined against reasonable baselines.
7. The impacts of joint implementation activities would have to be assessed with respect to their economic and social, as well as environmental, effects.
8. Joint implementation activities should, where appropriate, be accompanied by measures to ensure their long-term environmental benefits.
9. Joint implementation activities could address any greenhouse gas or any combination of gases.
10. Parties should give priority to joint implementation activities resulting in emissions limitations.
11. The benefits of joint implementation activities may be shared between the Parties involved.
12. Each of the Parties involved in a joint implementation activity would have to communicate relevant information thereon to the COP.

Source: Interim secretariat of the Framework Convention on Climate Change, 'Matters relating to commitments Criteria for Joint Implementation', in Kilaparti Ramakrishna (ed.), *Criteria for Joint Implementation under the Framework Convention on Climate Change*, Woods Hole Research Center, 1994, pp.51–8.

Draft set of Criteria for the 10th Session of the INC

A pilot phase shall be started to clarify the design of Joint Implementation projects and their costs/benefits. Moreover problems concerning groups of countries shall be considered. The preliminary criteria are to be checked and the institutional frame to be built. All contracting parties can participate. The pilot phase could start after the first Conference of the Parties or at a passed date to recognise ongoing projects. Its duration is not yet clear. Industrial countries investing in Joint Implementation shall inform about them in their national reports. Other countries have to deliver special reports. Countries could also deliver joint reports. In the pilot phase there is no crediting of emissions reduction. The following criteria shall apply:

A. Joint implementation is a voluntary activity under the responsibility of two or more parties; such activity must be undertaken or accepted by the Governments concerned.
B. Joint implementation should be beneficial to all Parties involved, and be consistent with their national priorities for sustainable development.
C. Joint implementation activities should bring about real and measurable results, determined against reasonable baselines.
D. The impacts of joint implementation activities would have to be assessed with respect to their economic and social, as well as environmental, effects.
E. Joint implementation activities should be accompanied by measures to ensure their long-term environmental benefits.

Concerning the institutional framework, the paper is vague. It discusses the project exchange and clearing house concept as well as an approach similar to the GEF. Furthermore, the Conference of the Parties shall have evaluation powers.

Source: UN, *Matters relating to commitments – Criteria for Joint Implementation*, A/AC.237/66, Geneva, 1994.

FRAMEWORK AGREEMENT BETWEEN THE USA AND COSTA RICA, 30 SEPTEMBER 1994

Statement of Intent for Bilateral Sustainable Development, Co-operation and Joint Implementation of Measures to reduce Emissions of Greenhouse Gases between the Government of the United States of America and the Government of the Republic of Costa Rica

Whereas, the government of the United States of America, as represented by the United States Department of Energy (DOE) and the United States Environmental Protection Agency (EPA), on the one hand; and the Government of the Republic of Costa Rica, as represented by the Costa Rica Ministry of Natural Resources, Energy and Mines, on the other hand, ("the Participants"), recognise that enhancing environmental protection, and in particular, controlling greenhouse gas emissions to limit potential adverse climate change impacts, would be mutually beneficial;
Whereas, the Participants recognise that limiting the adverse impacts of climate change requires a global solution, to which both the United States and Costa Rica can make significant contributions, and the Participants have a mutual interest in working together in this area;

Whereas, the Participants recognise that the Framework Convention on Climate Change, which the governments of both countries have ratified, and other international agreements including the Convention on Biological Diversity, to which the governments of both countries are signatories, encourage countries to pursue the rapid development of and joint implementation of co-operative, mutually voluntary, and cost-effective projects, particularly technology co-operation projects aimed at reducing or sequestering emissions of greenhouse gases and at promoting sustainable development and biological diversity conservation;
Whereas, the Participants will mutually benefit from the development and use of sustainable energy and greenhouse gas emission reduction and sequestration technologies and methods;
Whereas, the Participants recognise the potential for additional investment in environmentally, socially and economically sound development through the participation of the private sector in joint implementation of measures and technology co-operation projects to reduce emission of greenhouse gases;
Whereas, the Participants recognise that many methods and technologies that limit greenhouse gas emissions also contribute to the control of local and regional environmental problems and that cost-effective, world-wide greenhouse gas emission reductions may be achieved by encouraging such reductions in countries where responsive solutions are available at least cost with possible financial and technical assistance and investment from individuals and organisations in other industrialised countries;

The Participants declare as follows:
The Participants hereby intend to facilitate the development of joint implementation projects which will encourage the following: market development of greenhouse gas-reducing technologies, including energy efficiency and renewable energy technologies; education and training programs; increased diversification of energy sources; conservation, restoration, and enhancement of forest carbon sinks, especially in areas that promote biodiversity conservation and ecosystem protection; reduction of greenhouse gas emissions and other pollution; and the exchange of information regarding sustainable forestry and energy technologies;
This co-operation should provide the basis for future similar arrangements among countries, particularly in the Western Hemisphere, and contribute to the international establishment of an accessible joint implementation regime that is sensitive to environmental, developmental, social and economic priorities. This co-operation will encourage partnerships involving the Participants, the private sector, non-governmental organisations, and other entities.

The Participants intend that the forms of co-operation under this framework may include, but are not limited to, the following:

A. The design of Costa Rica's criteria to facilitate acceptance of joint implementation projects – including a model statement of acceptance – consistent with the Groundrules for the U.S. Initiative on Joint Implementation (USIJI) and Costa Rica's domestic priorities for measures to reduce greenhouse gas emissions and increase sinks via joint implementation, including biodiversity conservation and ecosystem protection, reduction of local pollution, sustainable land-use practices, improved rural income opportunities, and local participation in project planning and execution;

B. Identification and support of projects that are likely to meet the criteria for joint implementation project selection developed by the USIJI, by Costa Rica, and for the pilot phase adopted by the Conference of the Parties of the United Nations Framework Convention on Climate Change;

C. The design of methodologies and mechanisms to establish procedures for monitoring and external verification of greenhouse gas reductions, and the tracking and attribution of such reductions, consistent with the criteria for project selection being developed by the USIJI, Costa Rica and by the Conference of the Parties of the United Nations Framework Convention on Climate Change;

D. Outreach and promotion of joint implementation and other sustainable development activities in the private, public, and non-governmental sectors, including dissemination of information about the USIJI and about Costa Rica's criteria for joint implementation projects, and supporting technical assistance resources through workshops, conferences, and information networks;

E. Education and training activities for personnel in related energy and land-use sectors, both private and governmental, to strengthen human resources and institutional infrastructure and facilitate private joint venture activities;

F. Improvement of information networks to share information on technologies, potential partners, projects, technical resources, relevant policies, financing resources, and other information deemed useful to support joint implementation and sustainable development;

G. The promotion of the development of a hemispheric partnership on sustainable development and joint implementation projects which may be addressed at the Summit of the Americas in December 1994, and subsequent related activities;

H. The expansion of existing relevant intergovernmental initiatives for technical, economic, and scientific collaboration in the area of greenhouse gas emissions reduction;

I. The exploration of credible certification of emissions reductions, especially the determination of reasonable greenhouse gas emissions baselines at the project level, and its relevance to potential related market mechanisms;
J. The design of innovative financial arrangements for activities and projects implemented in accordance with this Statement of Intent, for the purposes of:
   1. facilitating increased private sector investment in sustainable development and joint implementation projects;
   2. promoting new partnerships for project financing;
   3. advocating on behalf of approved project portfolios for funding; and
   4. identifying additional sources of project funding and the policy framework needed to facilitate access to them, i.e., financial incentives.

The Participants intend to develop a Memorandum of Understanding to delineate further the intended co-operation activities listed above, as deemed necessary.

The Participants intend to examine the need for provisions to insure against loss of greenhouse gas emissions reductions achieved through jointly implemented projects.

The Participants intend that any joint implementation project or other activity or arrangement undertaken pursuant to this Statement of Intent will be on terms accepted by all parties to the transaction. Furthermore, the Participants intend to include appropriate patent and other intellectual property rights provisions, as well as provisions to protect business confidential information, in any such plans or arrangements. In particular, in the event that any activity involves access to and the sharing or transfer of technology subject to patents or other intellectual property rights, such access and sharing or transfer will be provided on terms which recognise and are consistent with the adequate and effective protection of intellectual property rights.

Washington, September 30, 1994

CRITERIA OF THE USIJI

Section I – Purpose

The purpose of the pilot program shall be to:

(1) encourage the rapid development and implementation of co-operative, mutually voluntary, cost-effective projects between U.S. and foreign partners aimed at reducing or sequestering emissions of greenhouse gases, particularly projects promoting technology co-operation with and sustainable development in developing countries and countries with economies in transition to market economies;

(2) promote a broad range of co-operative, mutually voluntary projects to test and evaluate methodologies for measuring, tracking and verifying costs and benefits;
(3) establish an empirical basis to contribute to the formulation of international criteria for joint implementation;
(4) encourage private sector investment and innovation in the development and dissemination of technologies for reducing or sequestering emissions of greenhouse gases; and
(5) encourage participating countries to adopt more complete climate action programs, including national inventories, baselines, policies and measures, and appropriate specific commitments.

Section II – Evaluation and Reassessment of Pilot Program

The pilot program shall be evaluated and reassessed within two years of its inception or within six months of adoption of international criteria for joint implementation by the Conference of the Parties to the United Nations Framework Convention on Climate Change, whichever is earlier.

Section III – Eligible Participants

A. Domestic

(1) Any U.S. citizen or resident alien;
(2) any company, organization or entity incorporated under or recognized by the laws of the United States, or group thereof; or
(3) any U.S. federal, state or local government entity.

B. Foreign

(1) Any country that has signed, ratified or acceded to the United Nations Framework Convention on Climate Change;
(2) any citizen or resident alien of a country identified in B(1) of this section;
(3) any company, organization or entity incorporated under or recognized by the laws of a country identified in B(1) of this section, or group thereof; or
(4) any national, provincial, state, or local government entity of a country identified in B(1) of this section.

Section IV – Evaluation Panel

A. An Evaluation Panel is hereby established.

B. The Evaluation Panel shall consist of eight members, of whom:

(1) one shall be an employee of the Department of Energy, who shall serve as Co-Chair;
(2) one shall be an employee of the Environmental Protection Agency, who shall serve as Co-Chair;
(3) one shall be an employee of the Agency for International Development;
(4) one shall be an employee of the Department of Agriculture;
(5) one shall be an employee of the Department of Commerce;

(6) one shall be an employee of the Department of the Interior;
(7) one shall be an employee of the Department of State; and
(8) one shall be an employee of the Department of the Treasury.

C. The Panel shall be responsible for:

(1) advising and assisting prospective U.S. and foreign participants on the technical parameters (including with respect to baselines, measuring and tracking) of projects submitted for inclusion in the USIJI;
(2) accepting project submissions from eligible U.S. participants and their foreign partners;
(3) reviewing and evaluating project submissions, including baseline projections;
(4) approving or rejecting project submissions for inclusion in the USIJI, based on criteria contained in section V;
(5) providing written reasons for its decisions, which shall be made publicly available, within 90 days of receipt of a complete submission or resubmission;
(6) certifying emissions reduced or sequestered estimated to result from projects;
(7) developing operational modalities for the implementation of the Program; and
(8) preparing an annual report of its activities, including a summary of approved projects.

Section V – Criteria

A. To be included in the USIJI, the Evaluation Panel must find that a project submission:

(1) is acceptable to the government of the host country;
(2) involves specific measures to reduce or sequester greenhouse gas emissions initiated as the result of the U.S. Initiative on Joint Implementation, or in reasonable anticipation thereof;
(3) provides data and methodological information sufficient to establish a baseline of current and future greenhouse gas emissions:
    (a) in the absence of the specific measures referred to in A.(2) of this section; and
    (b) as the result of the specific measures referred to in A.(2) of this section;
(4) will reduce or sequester greenhouse gas emissions beyond those referred to in A.(3)(a) of this section, and if federally funded, is or will be undertaken with funds in excess of those available for such activities in fiscal year 1993;
(5) contains adequate provisions for tracking the greenhouse gas emissions reduced or sequestered resulting from the project, and on a periodic

basis, for modifying such estimates and for comparing actual results with those originally projected;

(6) contains adequate provisions for external verification of the greenhouse gas emissions reduced or sequestered by the project;

(7) identifies any associated non-greenhouse gas environmental impacts/benefits;

(8) provides adequate assurance that greenhouse gas emissions reduced or sequestered over time will not be lost or reversed; and

(9) provides for annual reports to the Evaluation Panel on the emissions reduced or sequestered, and on the share of such emissions attributed to each of the participants, domestic and foreign, pursuant to the terms of voluntary agreements among project participants.

B. In determining whether to include projects under the USIJI, the Evaluation Panel shall also consider:

(1) the potential for the project to lead to changes in greenhouse gas emissions elsewhere;

(2) the potential positive and negative effects of the project apart from its effect on greenhouse gas emissions reduced or sequestered;

(3) whether the U.S. participants are emitters of greenhouse gases within the United States and, if so, whether they are taking measures to reduce or sequester such emissions; and

(4) whether efforts are underway within the host country to ratify or accede to the United Nations Framework Convention on Climate Change, to develop a national inventory and/or baseline of greenhouse gas emissions by sources and removals by sinks, and whether the host country is taking measures to reduce its emissions and enhance its sinks and reservoirs of greenhouse gases.

Source: U.S. Department of State: Announcement of Groundrules for U.S. Initiative on Joint Implementation, Washington.

BERLIN DECISION ON AIJ, 7 APRIL, 1995 (DECISION 5/CP.1)

The Conference of the Parties,

Recalling that, in accordance with Article 4.2(d) of the United Nations Framework Convention on Climate Change, the Conference is required to take decisions regarding criteria for joint implementation as indicated in Article 4.2(a),

Noting that the largest share of historical and current global emissions of greenhouse gases has originated in developed countries, that per capita emissions in developing countries are still relatively low and that the share of global emissions originating in developing countries will grow to meet their social and development needs,

Acknowledging that the global nature of climate change calls for the widest possible cooperation by all countries and their participation in an effective and appropriate international response, in accordance with their common but differentiated responsibilities and respective capabilities and their social and economic conditions,

Recognizing that,

(a) According to the provisions of the Convention, the commitments under Article 4.2(a) to adopt national policies and to take corresponding measures on the mitigation of climate change apply only to Parties included in Annex I to the Convention (Annex I Parties), and that Parties not included in Annex I to the Convention (non-Annex I Parties) have no such commitments,

(b) Activities implemented jointly between Annex I Parties and non-Annex I Parties will not be seen as fulfilment of current commitments of Annex I Parties under Article 4.2(b) of the Convention; but they could contribute to the achievement of the objective of the Convention and to the fulfilment of commitments of Annex II Parties under Article 4.5 of the Convention,

(c) Activities implemented jointly under the Convention are supplemental, and should only be treated as a subsidiary means of achieving the objective of the Convention,

(d) Activities implemented jointly in no way modify the commitments of each Party under the Convention,

1. *Decides:*

(a) To establish a pilot phase for activities implemented jointly among Annex I Parties and, on a voluntary basis, with non-Annex I Parties that so request;

(b) That activities implemented jointly should be compatible with and supportive of national environment and development priorities and strategies, contribute to cost-effectiveness in achieving global benefits and could be conducted in a comprehensive manner covering all relevant sources, sinks and reservoirs of greenhouse gases;

(c) That all activities implemented jointly under this pilot phase require prior acceptance, approval or endorsement by the Governments of the Parties participating in these activities;

(d) That activities implemented jointly should bring about real, measurable and long-term environmental benefits related to the mitigation of climate change that would not have occurred in the absence of such activities;

(e) That the financing of activities implemented jointly shall be additional to the financial obligations of Parties included in Annex II to the Convention within the framework of the financial mechanism as well as to current official development assistance (ODA) flows;

(f) That no credits shall accrue to any Party as a result of greenhouse gas emissions reduced or sequestered during the pilot phase from activities implemented jointly;

2. *Further decides* that during the pilot phase:

(a) The Subsidiary Body for Scientific and Technological Advice will, in coordination with the Subsidiary Body for Implementation, establish a framework for reporting, in a transparent, well-defined and credible fashion, on the possible global benefits and the national economic, social and environmental impacts as well as any practical experience gained or technical difficulties encountered in activities implemented jointly under the pilot phase;

(b) The Parties involved are encouraged to report to the Conference of the Parties through the secretariat using the framework thus established. This reporting shall be distinct from the national communications of Parties;

(c) The Subsidiary Body for Scientific and Technological Advice and the Subsidiary Body for Implementation, with the assistance of the secretariat are requested to prepare a synthesis report for consideration by the Conference of the Parties;

3. *Further decides*:

(a) That the Conference of the Parties shall, at its annual session, review the progress of the pilot phase on the basis of the synthesis report with a view to taking appropriate decisions on the continuation of the pilot phase;

(b) In so doing, the Conference of the Parties shall take into consideration the need for a comprehensive review of the pilot phase in order to take a conclusive decision on the pilot phase and the progression beyond that, no later than the end of the present decade.

KYOTO PROTOCOL FLEXIBLE INSTRUMENTS, 1997

Article 3

1. The Parties included in Annex I shall, individually or jointly, ensure that their aggregate anthropogenic carbon dioxide equivalent emissions of the greenhouse gases listed in Annex A do not exceed their assigned amounts, calculated pursuant to their quantified emission limitation and reduction commitments inscribed in Annex B and in accordance with the provisions of this Article, with a view to reducing their overall emissions of such gases by at least 5 per cent below 1990 levels in the commitment period 2008 to 2012.

[...]

7. In the first quantified emission limitation and reduction commitment period, from 2008 to 2012, the assigned amount for each Party included in Annex I shall be equal to the percentage inscribed for it in Annex B of its

aggregate anthropogenic carbon dioxide equivalent emissions of the greenhouse gases listed in Annex A in 1990 [...], multiplied by five.

[...]

10. Any emission reduction units, or any part of an assigned amount, which a Party acquires from another Party in accordance with the provisions of Article 6 or of Article 17 shall be added to the assigned amount for the acquiring Party.

11. Any emission reduction units, or any part of an assigned amount, which a Party transfers to another Party in accordance with the provisions of Article 6 or of Article 17 shall be subtracted from the assigned amount for the transferring Party.

[...]

13. Any certified emission reductions which a Party acquires from another Party in accordance with the provisions of Article 12 shall be added to the assigned amount for the acquiring Party.

Article 4

1. Any Parties included in Annex I that have reached an agreement to fulfil their commitments under Article 3 jointly, shall be deemed to have met those commitments provided that their total combined aggregate anthropogenic carbon dioxide equivalent emissions of the greenhouse gases listed in Annex A do not exceed their assigned amounts calculated pursuant to their quantified emission limitation and reduction commitments inscribed in Annex B and in accordance with the provisions of Article 3. The respective emission level allocated to each of the Parties to the agreement shall be set out in that agreement.

2. The Parties to any such agreement shall notify the secretariat of the terms of the agreement on the date of deposit of their instruments of ratification, acceptance or approval of this Protocol, or accession thereto. The secretariat shall in turn inform the Parties and signatories to the Convention of the terms of the agreement.

3. Any such agreement shall remain in operation for the duration of the commitment period specified in Article 3, paragraph 7.

4. If Parties acting jointly do so in the framework of, and together with, a regional economic integration organisation, any alteration in the composition of the organisation after adoption of this Protocol shall not affect existing commitments under this Protocol. Any alteration in the composition of the organisation shall only apply for the purposes of those commitments under Article 3 that are adopted subsequent to that alteration.

5. In the event of failure by the Parties to such an agreement to achieve their total combined level of emissions reductions, each Party to that agreement shall be responsible for its own level of emissions set out in the agreement.

6. If Parties acting jointly do so in the framework of, and together with, a regional economic integration organisation which is itself a Party to this Protocol, each member State of that regional economic integration organisation individually, and together with the regional economic integration organisation acting in accordance with Article 24, shall, in the event of failure to achieve the total combined level of emission reductions, be responsible for its level of emissions as notified in accordance with this Article.
[...]

Article 6
1. For the purpose of meeting its commitments under Article 3, any Party included in Annex I may transfer to, or acquire from, any other such Party emission reduction units resulting from projects aimed at reducing anthropogenic emissions by sources or enhancing anthropogenic removals by sinks of greenhouse gases in any sector of the economy, provided that:
(a) Any such project has the approval of the Parties involved;
(b) Any such project provides a reduction in emissions by sources, or an enhancement of removals by sinks, that is additional to any that would otherwise occur;
(c) It does not acquire any emission reduction units if it is not in compliance with its obligations under Articles 5 and 7; and
(d) The acquisition of emission reduction units shall be supplemental to domestic actions for the purposes of meeting commitments under Article 3.
2. The Conference of the Parties serving as the meeting of the Parties to this Protocol may, at its first session or as soon as practicable thereafter, further elaborate guidelines for the implementation of this Article, including for verification and reporting.
3. A Party included in Annex I may authorize legal entities to participate, under its responsibility, in actions leading to the generation, transfer or acquisition under this Article of emission reduction units.
4. If a question of implementation by a Party included in Annex I of the requirements referred to in this Article is identified in accordance with the relevant provisions of Article 8, transfers and acquisitions of emission reduction units may continue to be made after the question has been identified, provided that any such units may not be used by a Party to meet its commitments under Article 3 until any issue of compliance is resolved.
[...]

Article 12
1. A clean development mechanism is hereby defined.

2. The purpose of the clean development mechanism shall be to assist Parties not included in Annex I in achieving sustainable development and in contributing to the ultimate objective of the Convention, and to assist Parties included in Annex I in achieving compliance with their quantified emission limitation and reduction commitments under Article 3.
3. Under the clean development mechanism:
(a) Parties not included in Annex I will benefit from project activities resulting in certified emission reductions; and
(b) Parties included in Annex I may use the certified emission reductions accruing from such project activities to contribute to compliance with part of their quantified emission limitation and reduction commitments under Article 3, as determined by the Conference of the Parties serving as the meeting of the Parties to this Protocol.
4. The clean development mechanism shall be subject to the authority and guidance of the Conference of the Parties serving as the meeting of the Parties to this Protocol and be supervised by an executive board of the clean development mechanism.
5. Emission reductions resulting from each project activity shall be certified by operational entities to be designated by the Conference of the Parties serving as the meeting of the Parties to this Protocol, on the basis of:
(a) Voluntary participation approved by each Party involved;
(b) Real, measurable, and long-term benefits related to the mitigation of climate change; and
(c) Reductions in emissions that are additional to any that would occur in the absence of the certified project activity.
6. The clean development mechanism shall assist in arranging funding of certified project activities as necessary.
7. The Conference of the Parties serving as the meeting of the Parties to this Protocol shall, at its first session, elaborate modalities and procedures with the objective of ensuring transparency, efficiency and accountability through independent auditing and verification of project activities.
8. The Conference of the Parties serving as the meeting of the Parties to this Protocol shall ensure that a share of the proceeds from certified project activities is used to cover administrative expenses as well as to assist developing country Parties that are particularly vulnerable to the adverse effects of climate change to meet the costs of adaptation.
9. Participation under the clean development mechanism, including in activities mentioned in paragraph 3(a) above and in the acquisition of certified emission reductions, may involve private and/or public entities, and is to be subject to whatever guidance may be provided by the executive board of the clean development mechanism.

10. Certified emission reductions obtained during the period from the year 2000 up to the beginning of the first commitment period can be used to assist in achieving compliance in the first commitment period.

[...]

Article 17
The Conference of the Parties shall define the relevant principles, modalities, rules and guidelines, in particular for verification, reporting and accountability for emissions trading. The Parties included in Annex B may participate in emissions trading for the purposes of fulfilling their commitments under Article 3. Any such trading shall be supplemental to domestic actions for the purpose of meeting quantified emission limitation and reduction commitments under that Article.

The "Lost Article" on emission trading
1. For the purpose of meeting its commitments under Article 3, any Party included in Annex I may [, under the international framework to be established under paragraph 4 below,] transfer to or acquire from any other Party included in Annex I any part of its defined amount under Article 3, provided that each such Party is in compliance with its obligations under Articles [2, 3,] 5 and 8, and has in place a national mechanism for the certification and verification of emission trades.
2. A Party may authorise intermediaries to participate, under the responsibility of that Party, in actions leading to the transfer or acquisition, under this Article, of any part of its defined amount.
3. Emissions trading, as defined in paragraph 1 above, shall be subject to the following criteria:
[(a) Emission levels achieved before the start of any trading system established under this Protocol can[not] be used as the basis for emissions trading;]
(b) Emissions trading shall be supplemental to domestic policies and measures [, which should provide the main means] [for the purposes] of meeting commitments under Article 3; and
(c) A Party whose emissions are in excess of its defined amount in any budget period may acquire, but may not transfer, part of its defined amount.
4. The Conference of the Parties serving as the meeting of the Parties to this Protocol shall, at its first session or as soon as practicable thereafter, decide upon modalities, rules and guidelines for emissions trading, as provided for in paragraph 1 above, including methodologies for verification and reporting.
5. If a question of a Party's implementation of the requirements of Articles [2,3,] 5 or 8 is identified in accordance with the provisions of Article 9,

transfers and acquisitions of any part of a defined amount may continue to be made, provided that any such part of a defined amount may not be used by any Party to meet its obligations under Article 3 until any issue of compliance is resolved. If a question of a Party's implementation of paragraph 3(c) above is identified in accordance with the provisions of Article 9, the provisions of this paragraph shall apply only to transfers of any part of a defined amount by that Party.

Index